# Designing Mechanical Systems Using Autodesk® Building Systems

**Mary Ellen Pennisi-Vazzana**
**David Driver**

**autodesk®** Press

**THOMSON**

**DELMAR LEARNING**

Australia • Canada • Mexico • Singapore • Spain • United Kingdom • United States

**THOMSON**

**DELMAR LEARNING**

**autodesk** Press

# Designing Mechanical Systems Using Autodesk® Building Systems
## Mary Ellen Pennisi-Vazzana / David Driver

**Autodesk Press Staff**

**Vice President,
Technology and Trades SBU:**
Alar Elken

**Editorial Director:**
Sandy Clark

**Senior Acquisitions Editor:**
James DeVoe

**Senior Developmental Editor:**
John Fisher

**Marketing Director:**
Maura Theriault

**Channel Manager:**
Fair Huntoon

**Marketing Coordinator:**
Sarena Douglass

**Production Director:**
Mary Ellen Black

**Production Manager:**
Andrew Crouth

**Production Editor:**
Stacy Masucci

**Senior Art & Design Coordinator:**
Mary Beth Vought

**Editorial Assistant:**
Mary Ellen Martino

COPYRIGHT © 2003 Thomson Learning™.

Printed in Canada
1 2 3 4 5 XXX 06 05 04 03 02

For more information, contact Autodesk Press, 5 Maxwell Drive, Clifton Park, NY 12065-2919.

Or find us on the World Wide Web at www.autodeskpress.com

**Library of Congress Cataloging-in-Publication Data**
Pennisi-Vazzana, Mary Ellen.
  Designing Mechanical Systems Using Autodesk Building Systems/ Mary Ellen Pennisi-Vazzana.
      p. cm.
  ISBN 1-4018-3413-2
  1. Engineering graphics. 2. Mechanical desktop. 3. Engineering design--Data processing. 4. AutoCAD. I. Title.

  T353 .V39 2002
  620'.0042'0285—dc21
                    2002035983

**Notice To The Reader**

The publisher and author do not warrant or guarantee any of the products described herein or perform any independent analysis in connection with any of the product information contained herein. The publisher and author do not assume and expressly disclaim any obligation to obtain and include information other than that provided to it by the manufacturer.

The reader is expressly warned to consider and adopt all safety precautions that might be indicated by the activities described herein and to avoid all potential hazards. By following the instructions contained herein, the reader willingly assumes all risks in connection with such instructions.

The publisher and author make no representations or warranties of any kind, including but not limited to, the warranties of fitness for particular purpose or merchantibility, nor are any such representations implied with respect to the material set forth herein, and the publisher and author take no responsibility with respect to such material. The publisher and author shall not be liable for any special, consequential, or exemplary damages resulting, in whole or in part, from the readers' use of, or reliance upon, this material.

# CONTENTS

# Chapter 5 Reviewing Your Designs ............................................... 155

# Chapter 6 Communicating Your Design ....................................... 199

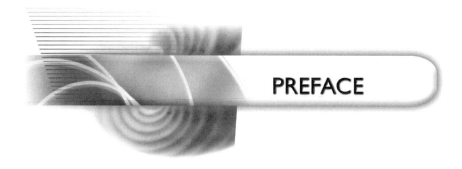

# PREFACE

## CHAPTER SUMMARIES

This book contains real-life drawing files that take you through the steps of the entire mechanical drafting design process, teaching you not only the features of Autodesk Building Systems, but also introducing you to the Autodesk Architectural Desktop features that aid you in the design process. Exposing you to features of Autodesk Architectural Desktop is necessary, primarily because most mechanical engineers are trained solely in AutoCAD, creating a steep learning curve to becoming comfortable with Autodesk Architectural Desktop. The lessons in this book assume no previous knowledge of Architectural Desktop.

### CHAPTER 1: THE BASICS

This chapter explains how to use the conventions contained within the book, and how to install and open the drawing files provided on the enclosed CD-ROM. This chapter highlights the Architectural Desktop and AutoCAD features that are necessary to understand in order to produce a mechanical plan.

Information in this chapter includes:

What You Should Know

Minimum System Requirements

How to Use This Book

Installing and Opening the Drawing Files

About Building Mechanical

Purchasing Building Mechanical

Shared Features

### CHAPTER 2: USING BUILDING SYSTEMS OBJECTS

This chapter introduces you to Building Systems objects, how they differ from Architectural Desktop objects and AutoCAD entities, and explains the fundamentals needed to create a mechanical system using Building Systems. The objective of this chapter is to provide you with a basic understanding of object functionality. This chapter contains primarily conceptual information.

Information in this chapter includes:

Introduction to Building Systems Objects

Fundamentals of AEC and Building Systems Objects

Creating Building Systems Objects

Connecting Building Systems Objects

Information That Is Stored Within Building Systems Objects

## CHAPTER 3: CREATING A MECHANICAL SCHEMATIC DIAGRAM

This chapter starts you out at a basic level, exposing you to the product's user interface, while simultaneously teaching you how to draft a schematic drawing. You learn how to place schematic lines to represent ducts and pipes. You then place schematic symbols to represent equipment and parts such as valves and dampers. This chapter also introduces the concept of assigning systems to the elements in your drawing. Once you have completed the hands-on tasks of drafting, annotating, and modifying a mechanical diagram, you will be introduced to some advanced product features that allow you to customize your drawing to suit your working environment.

Lessons in this chapter include:

Specifying Your Drawing Options

Drafting Your Mechanical Schematic Diagram

Modifying Your Mechanical Schematic Diagram

Creating Schematic Line Styles, System Definitions, and Symbol Styles

Annotating Your Mechanical Schematic Diagram

## CHAPTER 4: CREATING A MECHANICAL PLAN

Now that you have been exposed to the product's user interface you are ready for the more complex task of creating a mechanical plan. You will apply your knowledge of systems in order to place equipment and draft duct and pipe runs. This chapter also exposes you to drafting at different angles and elevations.

Lessons in this chapter include:

Specifying Your Drawing Options

Placing the Equipment

Working with Ducts and Fittings

Modifying Your Mechanical Plan

Working with System Definitions and Layer Keys

## CHAPTER 5: EXTRACTING INFORMATION FROM YOUR DRAWINGS

This chapter teaches you how to review the accuracy of your design. This chapter also introduces you to the Architectural Desktop functionality that enhances Building Systems. The Architectural Desktop features presented are key to extracting necessary information from the drawing.

Lessons in this chapter include:

Using Building Systems to Review Your Design

Checking Interference Using Sections

Scheduling MvParts

## CHAPTER 6: COMMUNICATING YOUR DESIGN

This chapter teaches you how to annotate the elements in their drawing. This chapter also covers different drawing sharing techniques in order to hand off your mechanical design to building contractors, suppliers, architects, electrical engineers, and fellow mechanical engineers and draftsmen.

Lessons in this chapter include:

Annotating Objects in Your Drawing

Annotating Your Mechanical Plan

Distributing Your Mechanical Design

## CHAPTER 7: CREATING CATALOG-BASED PARTS

This chapter takes you through the steps of content creation using the Content Builder and the Catalog Editor. This is an advanced chapter; however, you may need to create custom equipment if the equipment that you need is not included in Building Systems pre-created parts catalogs.

Lessons in this chapter include:

Working with the Catalog Editor

Working with Block-Based MvParts

Working with Parametric COLE-based MvParts

## APPENDIXES

The appendixes offer you supplemental information that is not directly associated with drafting a mechanical plan, but is useful for you to know in order to complete a project.

The appendixes include:

Appendix A: Calculating the Square Footage

Appendix B: Exploding Your Drawings

## ABOUT THE AUTHORS

**David Driver** has worked in architectural firms for the last 14 years, holding the positions of both project manager and computer manager. As project manager, David's experience covers a wide range of project types—from civic and commercial projects to single and multifamily housing. As CAD manager he has produced renderings and presentations, as well as performing the standard tasks of installation, configuration, training, and troubleshooting.

David is a professional architect, registered and practicing in the state of Arizona, working in collaboration with architects in Phoenix and Tucson. David provides software testing and training for Autodesk Architectural Desktop and Autodesk Building Systems. His continuing involvement in the construction industry ensures that his software training and consulting is highly relevant to his clients, while his involvement in the software industry provides his clients with in-depth knowledge and skill with computer-aided drafting and design.

**Mary Ellen Pennisi-Vazzana** has worked as a journalist and technical writer, and is AutoCAD Certified. Her technical writing for *Autodesk CAD Overlay* produced an Award of Excellence from the Society of Technical Communication (STC), an STC Award of Distinction, and an STC International Merit award. The CAD Overlay manual was also given a positive review in *Cadalyst* magazine. Mary Ellen has written AOTC courseware training for Autodesk Land Development Desktop and Autodesk Architectural Desktop.

Mary Ellen is the Lead Writer for the Autodesk Building Systems products. The *Autodesk Building Mechanical* and *Autodesk Building Electrical Release 1 Concepts Guides* both won a Distinguished award from the STC. The books also won the Best of Show award in the Northern New England STC Competition, and the *Building Mechanical Concepts Guide* went on to win an Award of Excellence in the International Competition.

## ACKNOWLEDGEMENTS

David and Mary Ellen would like to recognize and thank the following:

Walter Kappelman, for introducing David to technical writing and for convincing Mary Ellen to stay in it.

The Board of Challenger Learning Center of Alaska, housed inside of the Ted and Catherine Stevens Center for Science and Technology Education, in Kenai, Alaska, for permission to use this building in our sample files.

Special thanks to: Boyd Morgenthaler, P.E. of AMC Engineers (701 East Tudor Road, Suite 250 Anchorage, AK 99503) and Peter Klauder, President of Klauder & Company Associated Architects (606 Petersen Way Kenai, AK 99611) for granting permission to use their designs in this book, for without a building to work on, the sample files may never have been written.

We would also like to thank word warrior Karen Falkenstrom for her editorial eye, and Gregory A. Vazzana for his review and input on Content Builder.

Last (but not least) a thank you to the staff and contractors at Delmar Learning and Autodesk Press: James Devoe, John Fisher, Stacy Masucci, Mary Ellen Martino, and Mary Beth Vought of the Autodesk Press team; and Robert Shiery for performing a technical edit on the manuscript. Thanks also to the compositor, John Shanley of Phoenix Creative Graphics.

David acknowledges the following people:

My wife, Cantrell, for being patient and supportive while I created my parts of this book. Walter Kappelmann for providing me with my introduction to the world of technical writing. And Mary Ellen Vazzana for dragging me further into this world and keeping the schedule an important item in my life.

Mary Ellen acknowledges the following people:

Greg Vazzana, for his love and support. My parents, Anthony Pennisi and Elise Dubuque, and my brother, Michael. Also to my friends Debbie and John Santossuosso, who keep me laughing—even though the joke is always on me. And a final thanks to David Driver, whose genius and workaholism pulled this book together.

# CHAPTER 1

# The Basics

After you read this chapter, you should understand:

- the level of knowledge that this book requires
- the minimum system requirements of Autodesk® Building Systems
- how to use the lessons and exercises contained in this book
- what Building Systems enables you to do
- how to purchase Building Systems if you do not already own it
- which features of Autodesk® Architectural Desktop and AutoCAD are also used when working with Building Systems

## WHAT YOU SHOULD KNOW

This book requires some knowledge of AutoCAD. This includes the ability to draft simple line drawings using different snap options. This book also requires that you understand how to use some basic AutoCAD commands such as MOVE, STRETCH, ERASE, COPY, and PASTE.

This book assumes no knowledge of Autodesk Architectural Desktop. The book explains when an Architectural Desktop feature is used to accomplish a task. This explanation includes why it is necessary to use the feature, where to access the feature, and how to use the feature in conjunction with Building Systems. However, this book does not intend to teach Autodesk Architectural Desktop in its entirety, it only teaches you how to use the features necessary to complete a mechanical drawing. It would be beneficial to read through the "Shared Features" section of this chapter before stepping through the lessons. To learn more about Architectural Desktop, see *Mastering Autodesk® Architectural Desktop* by Paul F. Aubin and *Accessing Architectural Desktop* by William Wyatt.

This book assumes no knowledge of Building Systems. Every chapter contains the conceptual information necessary to complete the lessons. If you read this book cover to cover you should have a solid understanding of the Mechanical portion of the Building Systems product. If you choose to perform specific lessons that are not in the context or order of the book, then you will only know how to perform those specific tasks. However, this may be all that you need to do in order to produce your work-related deliverables. It is suggested that you read the "Chapter Descriptions" section in the preface to decide what chapters are relevant to performing your job tasks.

## MINIMUM SYSTEM REQUIREMENTS

In order to use the CD and lessons contained in this book, you must have access to a computer that has Building Systems Release 3 installed.

Minimum system requirements:

1. Operating systems
   - Windows® NT 4.0 with Service Pack 5.0
   - Windows 98 Second Edition
   - Windows Millennium Edition (ME)
   - Windows 2000 Professional
   - Windows XP
   - Windows 95 is **not** supported

2. Processor
   - Intel® Pentium® II or AMD-K6® based PC, with 450MHz processor (minimum)
   - Intel Pentium III or AMD-K6 III based PC, with 750MHz processor (recommended)

3. RAM
   - 128 MB (minimum)
   - 256 MB (recommended)

4. Video
   - 800 x 600 VGA with 8 MB video RAM (minimum)
   - 1024 x 768 SVGA with 32 MB video RAM (recommended)
   - Requires a Windows-supported display adapter

5. Hard disk
   - Installation 200 MB (minimum)
   - Swap space 64 MB (minimum)
   - System folder 60 MB (minimum)
   - 75 MB (recommended)
   - Shared files 40 MB

6. CD-ROM
   - For installation of Building Systems Release 3, as well as the drawing files contained on the CD-ROM of this book.

7. Hardware (optional)
   - Open GL-compatible 3D video card
   - Printer or plotter
   - Digitizer
   - Network interface card
   - The OpenGL driver that comes with the 3D graphics card must have the full support of OpenGL

8. Modem or access to an Internet connection
   - Microsoft Internet Explorer 5.0
   - Netscape Navigator 4.5 or later

 **Note:** Internet Explorer 5.5 is installed with Building Systems, because we will be using the Internet in Chapter 6, "Communicating Your Design".

## HOW TO USE THIS BOOK

This book contains seven chapters. Each chapter has one objective, which is represented in the chapter title. For example, "Reviewing Your Designs" contains a lesson that describes how to create a schedule by extracting information from a drawing file. The chapters contain lessons that teach you how to complete each task in order to meet the chapter objective. The lessons in the chapter are broken down into one or more exercises. Most exercises have a corresponding drawing file that is located on the CD-ROM included with this book.

### Installing and Opening the Drawing Files

When an exercise requires you to open a drawing file from the CD-ROM, you will see the following icon with the drawing name that you need to open beside it:

 **Open:** *03 Placing Schematic Symbols.dwg*

Before starting any lessons, make sure that Building Systems is loaded on top of Architectural Desktop by running a Building Systems command such as Mechanical ➤ Duct ➤ Add.

1. Access the drawing files on the CD-ROM:
   a. Place the CD in your CD-ROM drive.
   b. In the Autodesk Press dialog box, click the Drawing Files icon.
   c. Windows Explorer opens to a folder named Drawing Files, with a list of chapter folders beneath.

2. Open a specific drawing file:
   a. Browse to the chapter folder that you are working.
   b. Open the chapter folder to view a list of the drawing files.
   c. Click on the drawing file that the lesson asks you to open.

## ABOUT BUILDING SYSTEMS RELEASE 3

Autodesk Building Systems R2 consisted of two separate products - Building Mechanical and Building Electrical. Both products were add-ons to Autodesk Architectural Desktop. With Release 3 Autodesk combined these two products together with Architectural Desktop, and added plumbing into the mix. One install provides the Architectural Desktop base, as well as the mechanical, electrical and plumbing functionality. The mechanical portion of the Building Systems program enables you to design a complete building mechanical system by integrating functionality with Architectural Desktop. This integrated functionality is described in detail in the "Shared Features" section of this chapter.

You design a complete Building Systems system by creating system definitions, drafting ductwork and piping, modifying the ductwork and piping, annotating your drawing, and producing a schedule of the equipment in your drawing. Once your Building Systems system is complete, you are able to share your drawing with contractors, drafters, architects, and engineers by not only plotting your drawing, but by posting your drawing on the Internet so that others can access your drawing from remote locations. You are able to produce mechanical systems for residential, commercial, and industrial buildings.

When drafting in Building Systems, you can produce a two-dimensional (2D) schematic, plan, or isometric drawing, as well as a three-dimensional (3D) drawing that enables you to view the different elevation values contained in your drawing.

Building Systems contains catalogs of predefined equipment, parts, and accessories that are specific to the mechanical industry. As you draft your duct and pipe runs, Building Systems places the appropriate fittings at the turns and junctions of your runs. The Building Systems objects contained in the catalogs have predefined styles, sizes, shapes, and connectors. Each connector has a connector type that has a built-in domain, description, and connection points. Because of this, the Building Systems objects in your drawing are able to recognize one another by respecting what type of duct or pipe system style should connect with which connection points on a piece of equipment.

The following illustration shows a water source heat pump MvPart that has connectors for both ducts and pipes. The MvPart recognizes whether a duct or pipe should be connected to a specific connector and also has the ability to recognize the systems assigned to the duct and pipes.

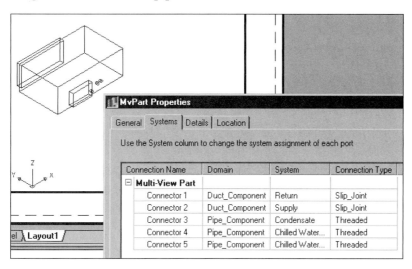

You also have the ability to create custom equipment and fittings, if the one that you need is not available within the predefined catalog of parts that are installed with Building Systems. Because Building Systems objects are such a fundamental part of the product, they are described in detail in Chapter 2, "Using Building Systems Objects".

## PURCHASING BUILDING SYSTEMS

If you do not already own Autodesk Building Systems, you can call 1-800-964-6432 for the contact information of the authorized reseller nearest to you. You can also access the Autodesk Resellers Web site at *www.autodesk.com*.

## SHARED FEATURES

Building Systems is built on Architectural Desktop, and therefore, integrates some Architectural Desktop functionality in order to complete the design process. Architectural Desktop is built on AutoCAD, therefore integrating AutoCAD functionality. However, you do not need to be a power AutoCAD user, or a knowledgeable Architectural Desktop user to be able to master Building Systems. The following information outlines the features of both Architectural Desktop and AutoCAD that Building Systems uses in order to produce a complete mechanical plan. There are many other features of AutoCAD and Architectural Desktop that you will never need to use in conjunction with Building Systems, so rather then attempting to learn AutoCAD and Architectural Desktop, focus on learning these specific features.

The following information provides a conceptual overview of these features, so that you can understand their relevance to Building Systems. The following sections do not attempt to teach you how to use these features; however, these features are integrated within the lessons of this book so that you can learn how to use them with relevance to a particular task. The following sections include the way to access the feature; it is recommended that you open them and become familiar with the feature's user interface. Use the online Help included with Architectural Desktop to help you.

### ANCHORS

Anchors are features that were initially introduced by Architectural Desktop. Anchors create a link between AEC objects and other drawing elements. Building Systems also has introduced anchoring features. There are several different anchoring commands available both in Architectural Desktop and Building Systems, but before the specifics of those commands are explained, it is important to understand the underlying objective of anchors, and why they are useful.

Creating an anchor, or link, between an AEC object and another drawing element binds the elements together. This 'bind' allows you to make updates and changes to the anchored elements simultaneously. For example, if you anchor an air-handling unit to a duct, and then move the duct, the air-handling unit will move with it. If you erase the duct, then the air-handling unit that is anchored to it is also erased. Therefore, creating an anchor relationship between drawing objects makes updating your drawing easier.

Architectural Desktop has integrated some of this functionality into the program, so that anchoring relationships are created automatically, without you needing to do anything. For example, doors are anchored to walls. So when you place a door in your drawing, the door 'knows' that it is part of the wall. Therefore, if the wall is moved, the door moves as well.

Building Systems has also integrated some automatic anchoring functionality into the program. For example, when you place a schematic symbol onto a schematic line. The schematic symbol is anchored to the line.

You can compare anchoring objects to creating a 'parent to child' relationship. Whatever you are anchoring *to* becomes the parent in the relationship. For example, if you anchor an air-handling unit *to* a duct, then the duct becomes the parent. Any modifications that you make to the duct will be made to the air-handling unit as well. However, if you make modifications to the air-handling unit that is anchored to the duct, then no changes are made to the duct. In this circumstance, the duct is the parent, and the air-handling unit is the child.

 **Note:** Anchoring an air-handling unit to a duct changes the orientation of the air-handling unit to be perpendicular to the duct. You need to manually change the orientation back to the desired position.

There are three Building Systems anchoring commands:

1. **Building Systems Curve Anchors** are used to link any AEC object with any other AEC object. This means that you can use it to link a Building Systems object such as a pipe, to an Architectural Desktop object, such as a wall. You can also use this command to link two Building Systems objects, such as a Pipe to a Heating Water Unit.

   You can access Building Systems Curve Anchors from the MEP Common menu, by clicking Anchors ➤ Building Systems Curve Anchors.

2. **Building Systems System Anchors** are used to link an object to a point along a linear object like a duct or pipe.

   You can access Building Systems System Anchors from the MEP Common menu, by clicking Anchors ➤ Building Systems System Anchors.

3. **Building System Reference Anchors** are used to anchor linear elements between two other objects such as a duct between two fittings.

### ANNOTATION

Annotating your drawing is an important step in producing construction documents and provides clarification on the different drawing aspects in your design.

Building Systems works in conjunction with Architectural Desktop to supply you with the annotating features that you need. Building Systems offers labeling commands that automatically label ducts and pipes. Labels are style-based and are able to read the information stored in an object. Architectural Desktop supplies you with documentation symbols such as title marks, leaders, and detail marks.

To annotate a drawing you use labeling commands included with Building Systems from the MEP Common menu, by clicking on Labels. To access Architectural Desktop's documentation symbols, click Documentation ➤ Documentation Content, and select from the list of documentation content available. Selecting a piece of documentation content opens the DesignCenter. Refer to the "DesignCenter" section for more information.

## AREAS AND AREA GROUPS

Areas are an Architectural Desktop feature that represents a two-dimensional (2D) room, such as a mechanical room, and is used to calculate space. Once areas are defined in your drawing, you can create area groups to differentiate areas contained in your drawing. You use these commands to calculate the square footage of your mechanical plan. For example, once the area and area groups are drawn, you can group the areas in your drawing into zones and then export the area calculations into an Excel® spreadsheet. To access the area commands, from the Documentation menu click Areas or Area Groups.

## CEILING GRIDS

Architectural Desktop has features that allow you to create ceiling grids that contain cells to place terminals and diffusers. You may receive a drawing that already has a ceiling grid drawn, or you may want to draw the ceiling grid yourself. The ceiling grid may already contain an elevation value, however, the mechanical parts that you place on the ceiling grid do not automatically inherit that same elevation value. If you start a duct or pipe run at a ceiling terminal, then you need to specify the duct or pipe elevation to be higher then the elevation of the ceiling terminal.

Ceiling grids contain a hidden NODE at each of the intersections. You can set your snap to detect the nodes and use them as a guide as you place MvParts and schematic symbols. However, the height of the ceiling grid determines the $Z$ elevation of the MvPart or schematic symbol.

To access the ceiling grid commands, from the MEP Commom menu click Grids ➤ Add Ceiling Grids.

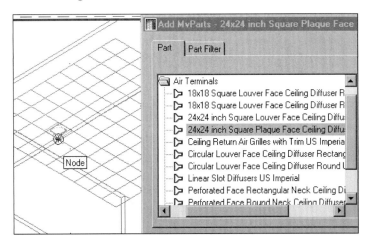

## DESIGNCENTER

DesignCenter is an AutoCAD feature that has integrated Architectural Desktop symbols and styles. DesignCenter provides a way to reuse information from drawings, blocks, or external references (xref's) without needing to leave your drawing session to access them. You can transfer information such as layer definitions, linetypes, layout tabs, text and dimension styles, or customized drawing content by dragging and dropping into open drawing files, network drives, or Internet locations. You can also right-click shortcut menus for managing, inserting, copying, and opening drawings.

The DesignCenter is broken down into two panels. The left panel is the navigation pane, or tree view, this is a hierarchical listing of drawing files or content folders. The right panel contains the content pane, or palette. This panel shows the drawing file content. The right panel also contains a visual preview of the selected drawing content. This information may be viewed from any direction by dragging over the view when the orbit icon appears.

You can access specific folders within DesignCenter from the associated menu command. For example, if you wanted to add a title mark, then from the Documentation menu click Documentation Content ➤ Title Marks.

## DISPLAY MANAGER

Architectural Desktop and Building Systems contain AEC objects that you can use when drawing your mechanical system. AEC objects can be both three-dimensional (3D), and schematic, and can be viewed in many different orientations. There are several view orientations associated with each AEC object, and each object can contain different view orientations, depending on the specific views necessary for the object. There is also the ability to assign different view orientations to an entire drawing by using layout tabs, such as Model. The functionality associated with all of the different view orientations for objects and drawings in Architectural Desktop is called the *display system*. All display system functionality is accessed through the Display Manager.

The importance of using the Display Manager is the ability to control how a drawing appears. This ability enables you to view the same drawing in different ways. For example, you may want to view your drawing as a one-line schematic in order to view all of the different drawing elements on a flat plane. However, using Display Manager you can also view the same drawing as a riser diagram by using an isometric view, in order to see the flow direction and slope of the mechanical layout.

The *display systems*, or view orientations of an object or drawing can be created, modified, and reused through the Display Manager. The Display Manager is used throughout the lessons in this book, and is an essential feature to understand when drafting in Building Systems.

The Display Manager is broken down into two panels. The left panel allows you to browse available display configurations, display sets, and display representations by object while the right panel displays the associated view representations within the selected folder. You can access Display Manager from the Desktop menu by clicking Display Manager.

Use the left panel
to navigate to
different display
system components

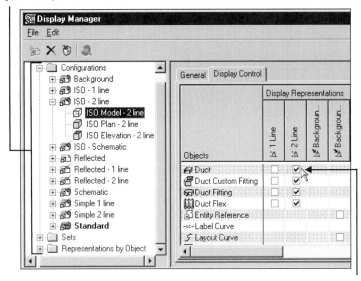

Use the right panel
to select and view
the different display
representations
available for each
object

## LAYER KEYS AND STANDARDS

Layer keys are a feature of Building Systems that work in conjunction with Architectural Desktop's Layer Manager. Each Building Systems Object has a layer key built into it. The Function of the layer key is to hold values for layers. As you draw a duct, Building Systems uses the layer key specified by the system of the duct to place the duct on. If the duct system's layer key points to a layer that does not exist yet in the drawing, the layer will automatically be created. With the MvParts, the layer key is held in the catalog. When the MvPart is placed in the drawing, like the duct, the layer key is read, and the part is placed on the layer pointed to by its layer key. The layer keys are already set up for you. Layer keys are grouped together in layer key styles. Layer key styles work similarly to Layer Manager in that each layer key style contains layer properties such as layer name, description, color, linetype, and lineweight.

Layer key styles are accessed through Style Manager under the layer key style type. Building Systems provides layer key styles that are based on industry layer standards, such as SMACNA. Using Style Manager you can create, edit, and delete layer keys.

To access the layer key styles contained in Building Systems, click the Desktop menu, and then click Layer Management ➤ Layer Key Styles. In Style Manager expand the Layer Key Styles folder, select the Mechanical layer key style, right-click, and click Edit. In the Layer Key Style Properties dialog box, click the Keys tab.

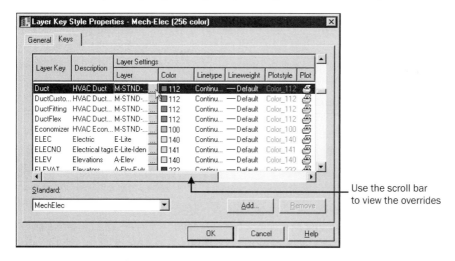

Use the scroll bar to view the overrides

You can click the ellipses button to access the Layer Name dialog box to edit specific properties such as the layer name, value, and description.

**Note:** There are 5 different layer keys provided for you to use. The default on install is the Mech-Elec. This layer key style contains all the layer keys for all objects in Building Systems. Likewise the Elec-Mech layer key style contains layer keys for all objects in Building Systems. The difference is in the features that are common such as labels and schematic lines and symbols. The Mech-Elec will place common items on a layer with a M-xxxx-xxxx, whereas the Elec-Mech will place common items on a layer with an E prefix.

## LAYER MANAGER

The Layer Manager originated in AutoCAD and was enhanced by Architectural Desktop and is useful in Building Systems. You use the Layer Manager to organize, view, and change the layer properties contained in your drawing. You can do this by making a layer that you are working on the current layer, creating new layers, sorting layers, and grouping layers. You can also create layering schemes so that you can work consistently using a desired layer configuration.

The difference between the Layer Properties Manager in AutoCAD and the Layer Manager accessed through Architectural Desktop are the icons across the top of the dialog box that access Layer Standards, Layer Key Styles, and Layer Overrides. Additionally, the Layer Manager gives you snapshots and layer groups to enable you to save and control a group of layers.

You access the Layer Manager from the Desktop menu by clicking Layer Management ➤ Layer Manager.

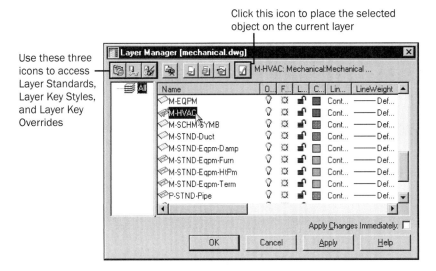

Click this icon to place the selected object on the current layer

Use these three icons to access Layer Standards, Layer Key Styles, and Layer Key Overrides

## LAYOUT CURVES

Layout curves work in conjunction with anchors. A layout curve is *not* an object or a polyline, but a means in which to place objects along a "pathway". You define the "pathway" or layout curve by selecting AEC objects, such as ducts, and then choosing the spacing in which you would like the duct to be placed along the layout curve. You are prompted to place the nodes to anchor the objects to in three ways:

1. Manual – where you place the nodes by selecting them on screen.
2. Repeat – to have the nodes equally spaced along the layout curve. If you choose this option, then the amount of nodes will change while maintaining a fixed distance between the nodes when you modify the length of the layout curve.

3.  Space Evenly – to specify the number of nodes that you would like along the layout curve. If you choose this option, then the amount of nodes stays the same if you modify the length of the layout curve, but the spacing will adjust accordingly.

## SCHEDULES

Schedules are a feature of Architectural Desktop that enable you to produce a detailed list of all of the objects in your drawing. The information that you decide to include in your schedule can be placed into a schedule table, or exported into a Microsoft® Excel spreadsheet. Any changes that you make in your drawing are automatically reflected in your schedule table.

You can create schedule table styles to specify what objects you want in your schedule table, and what specific object information that you would like to include in your schedule table. You can schedule properties of the Building Systems objects, or you can create properties manually and then input the values reported in the schedule.

**Note:** There are some sample schedule table styles contained in the program. You access the schedule table styles from the Desktop menu by clicking Style Manager. When in Style Manager, click Open Drawing from the File menu. Browse to the *Program Files\Autodesk Building Systems 3\Content\Imperial\Schedules\ABM Duct Schedule Tables (Imperial) 3.dwg* file.

To create a schedule, you need to define schedule data and attach this data to the objects and object styles in your drawing. You do this by defining property set definitions. Property set definitions are contained within an object, such as size information. For example, an MvPart property definition could include the name, a description, and a size value. Any property set definitions included with Building Systems are based on common industry schedule standards.

You can create new property set definitions, copy them into different drawings, or delete them if they are not necessary. You access Property Set Definitions through Style Manager. Open Style Manager from the Desktop menu by clicking Style Manager. In the right panel, browse to Property Set Definitions. Expand Property Set Definitions and select the specific property set definition that you would like to review.

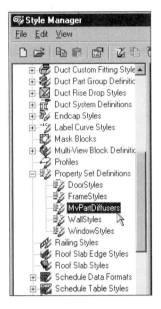

Once you have selected the property set, right-click and click Edit. Click the Definitions tab to view the list of available property definitions for that object. This feature is covered in detail in Chapter 5, "Reviewing Your Designs".

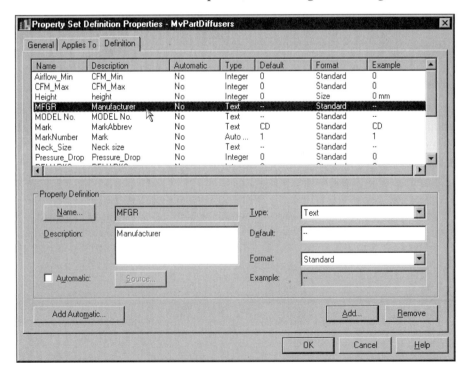

## SCHEDULE TAGS

Schedule tags are a feature of Architectural Desktop that work in conjunction with the scheduling feature. Schedule tags are annotation tools that provide you with the ability to link a piece of schedule data (a property in a property set definition) to a visible attribute tag. This way if you change the object, the schedule tag updates automatically.

The tags that are supplied with the software are created with a special format to the attribute that is buried in the tag. These attributes have the format:

```
PropertySetName:Property
```

The object tags supplied with Building Systems are created to import the property sets that they need as you place them. The property set definitions are not in the template files, however, they are automatically imported from the *AecbPropertySetDefs3.dwg*. This file is located under *Program Files\Autodesk Building Systems 3\Content\Imperial\Schedules\AecbPropertySetDefs3.dwg*.

You must use the naming convention; "PropertySetName:Property," for schedule tags to update dynamically with changes. For example, MvParts-AirTerminalObjects: CFM-Act.

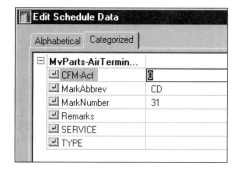

You cannot use spaces or extended names for your property set definitions. This causes the link between the schedule table and the schedule tag to break.

You can access schedule tags from the Documentation menu by clicking Schedule Tags ➤ Object Tags.

## SECTION LINES AND 2D SECTION OBJECTS

Sections are Architectural Desktop features that have been adopted by Building Systems. You use 2D section lines to create 2D section objects. The purpose of creating a 2D section object is to view where the different Building Systems objects are in relation to one another. Once the 2D section object is created, you can use different drawing views, such as isometric and plan, to review your drawing. This is an important step in identifying any places where objects may obstruct with one another. Examples of when a 2D section may be useful would be for mechanical rooms, HVAC layouts, and roof or wall penetrations.

When using this feature, be sure that all of the layers are turned on for the objects that you would like to view. You can also use Building Systems's two-line edge toggle command to hide unwanted edges when viewing two-line displays. If you want to add depth to your 2D section object you can create 'subdivisions' within your section and apply different lineweights to each subdivision.

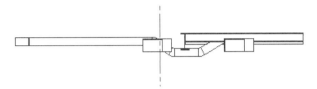

A 2D section object is linked to your drawing, and therefore can be updated when drawing changes are made. Access this feature from the Documentation menu by clicking Sections ➤ Add Section Line.

## SPACES

Spaces are an Architectural Desktop feature that is useful when working with Building Systems. Spaces are objects that represent, for example, rooms. Space objects have a built-in volume that includes area, length, width, and height. The importance of space objects in Building Systems is the ability to produce an area calculation. Space objects are also useful because they respect the distance above the ceiling to the top of the volume of the space. This makes it possible to calculate not only the size of the room, but the space needed between the ceiling and the roof or the floor above when laying out ductwork above a ceiling.

You can also create space styles to define what properties you want to include. These properties include the minimum and maximum area, length, and width. Once the maximum area is defined in the style, the program does not allow you to accidentally draw a space larger than the maximum size set by the style. Space styles can also be created to represent different zones in your drawing, such as the southern exposure, the core, and the northern exposures of an office building. Space objects have a built-in query function that enables you to get the square footage of the spaces based on the space style, or on each individual space.

You access spaces by entering **space** at the command line.

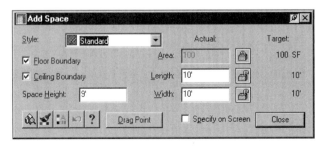

## STYLE MANAGER

Architectural Desktop and Building Systems contain styles that you can use when drawing your mechanical system. All AEC objects are style-based. Styles are a storage place for the variables associated with each of the different AEC objects. An AutoCAD analogy would be dimension styles and text styles. In addition to common variables, styles contain information on how the objects are to be displayed.

The style of an object can be created, modified, and reused through Style Manager. Style Manager allows you to copy styles between drawings. If you have two or more drawings open then you can drag and drop styles between drawings. Style Manager is used throughout the lessons in this book, and is an essential feature to understand when drafting in Building Systems.

Style Manager is broken down into two panels. The left panel allows you to browse available styles by object, while the right panel displays a description of the selected style. If you double-click on a selected style in the right panel, you access the associated Style Properties dialog box. The Style Properties dialog box enables you to change values of a style, such as the style's name, description, or display properties. Due to the fact that every object is different, you will see that the tabs and options contained in the Style Properties dialog box change, depending on the style that you are editing. You can access Style Manager from the Desktop menu by clicking Style Manager.

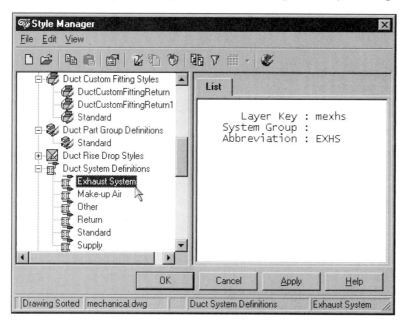

# KEY CONCEPTS: THE BASICS

This book requires some knowledge of AutoCAD, including how to use some basic AutoCAD commands such as MOVE, STRETCH, ERASE, COPY, and PASTE.

This book assumes no knowledge of Architectural Desktop or Building Systems.

Building Systems is a stand alone product built on AutoCAD2002 and Architectural Desktop 3.3.

When drafting in Building Systems, you can produce a two-dimensional (2D) schematic, plan, or isometric drawing, as well as a three-dimensional (3D) drawing.

Building Systems contains catalogs of pre-defined equipment, parts, and accessories that are specific to the mechanical industry.

Anchors create a link between AEC objects and other drawing elements.

Areas are an Architectural Desktop feature that represent a 2D room, such as a mechanical room, and are used to calculate space.

DesignCenter provides a way to reuse information from drawings, blocks, or external references (xrefs) without leaving your drawing session to access them.

The Display Manager controls how a drawing appears. This ability enables you to view the same drawing in different ways.

Layer keys are a feature of Architectural Desktop that provide automatic layer creation. Each Building Systems Object has a layer key built into it. The function of the layer key is to hold values for layers.

You use the Layer Manager to organize, view, and change the layer properties contained in your drawing.

The difference between the Layer Properties Manager in AutoCAD and the Layer Manager accessed through Architectural Desktop are the icons across the top of the dialog box that access Layer Standards, Layer Key Styles, and Layer Overrides.

Layout curves work in conjunction with anchors. A layout curve is *not* an object or a polyline, but a means in which to place objects along a "pathway".

Schedules are a feature of Architectural Desktop that enable you to produce a detailed list of all of the objects in your drawing.

Schedule tags are a feature of Architectural Desktop that work in conjunction with the scheduling feature. Schedule tags are annotation tools that provide you with the ability to link a piece of schedule data to a visible attribute tag.

You use 2D section lines to create 2D section objects in order to view where the different Building Systems objects are in relation to one another.

All AEC objects are style-based. Styles are accessed through Style Manager and are a storage place for the variables associated with each of the different AEC objects.

# CHAPTER 2

# Understanding AEC Objects

After you read this chapter, you should understand:

- what AEC objects are and how they differ from standard AutoCAD entities
- how to display the same AEC object in different views
- what styles are and how they relate to AEC objects
- how Architectural Desktop objects differ from Building Systems objects
- Building Systems objects
- how Building Systems objects relay information to each other through connection points
- what systems are and the importance of using them when drafting your mechanical design

## AEC OBJECTS VERSUS AUTOCAD ENTITIES

Architectural Desktop and Building Systems objects are called AEC objects. These AEC objects are specific to the architectural, engineering, and construction industries. They are the heart of Architectural Desktop and Building Systems because of their vast functionality. They are not only symbolic representations of an object, but they *are* the object. A pipe *knows* that is it a pipe, and therefore does not allow itself to be connected to a duct. However the functionality spans far beyond that, from how you can display AEC objects, to how AEC objects respect and relate to one another, and how you can use and create different styles of the same AEC object to customize your work environment. Since AEC objects know what they are, they also make scheduling and annotating a drawing more dynamic. These AEC objects are the key to creating a building model that is easily updated and maintained throughout the life cycle of the building. This fundamental functionality is explained in detail throughout this chapter, and is essential to understanding both Architectural Desktop and Building Systems. It is important to note that there are differences between the AEC objects contained in Architectural Desktop and the AEC objects contained in Building Systems. Examples of AEC objects contained in Architectural Desktop are doors, walls, and windows. Examples of AEC objects contained in Building Systems are ducts, pipes, and MvParts. These differences will also be explained in detail in this chapter.

AutoCAD entities are generic drawing objects such as lines, polylines, arcs, and circles. They do not know what they are. A line does not know that it is a line, nor should it. AutoCAD entities are flexible, so you can build objects that represent elements in your drawing. You could draw four lines in a rectangular pattern and call it a door. However, if you hand off your drawing to an external contractor, that person may not know that those four lines are meant to represent a door. Unlike a door object in Architectural Desktop, the four-line door will not contain the ability to display in different views, nor will it be anchored to a wall and move when the wall is moved. The important fact to remember is that AutoCAD entities are drawing elements that are not domain specific, and do not have the ability to recognize what they are.

The following illustration shows an example of an AutoCAD circle entity and a Building Systems base-mounted pump. You *could* draw the same base-mounted pump out of AutoCAD entities, but imagine the time that it would take. Then realize that the base-mounted pump created out of AutoCAD entities could never be dynamically updated, viewed in different directions, or connected to other drawing objects.

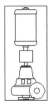

There are four qualities that differentiate all Building Systems objects from AutoCAD entities:

1.  AEC objects contain built-in instructions that tell the screen to "show the object this way".

    Picking two points on the screen creates a Building Systems duct segment.

    A set of values held in the object determine not only the color and linetype is drawn, but how the duct appears or is displayed (whether the duct is displayed as one line, two lines, or as a three-dimensional duct).

2.  AEC objects are able to relate to other objects.

    In Building Systems this relationship may be an anchor, like those established when a schematic symbol is placed on a schematic line. This relationship could also be a connector at the end of a duct. The connector passes information from one duct to another, and also enables you to resize just one duct to affect a whole branch of the system.

3.  AEC objects have styles.

    Most AEC objects are style-based. A change to the style affects all objects of that style. This can be compared to the behavior of AutoCAD text and dimension styles. However, AEC objects store much more information within a style than standard AutoCAD text and dimension styles.

4.  Building Systems objects use a System Definition.

    The System Definition is used to override the layering, provide a system label value, and to adjust the display color or linetype for all ducts and fittings that belong to the same system.

## HOW AEC OBJECTS DISPLAY IN DIFFERENT VIEWS

AEC objects contain display representations. Display representations control how different views of an AEC object appear in your drawing, allowing you to view the same AEC object in a 2D, 3D, or isometric view. Depending on the display needs of the object, the object can have one or several display representations associated with it. For example, a label curve only has one display representation, but an MvPart (multi-view part) such as a base-mounted pump has five display representations. Display representations are managed through Architectural Desktop's Display Manager.

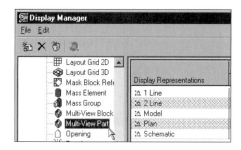

Notice that the five display representations associated with the MvPart are:

1. **1 Line** to view the object's 2D or 3D centerline, as well as viewing the object in an isometric view

2. **2 Line** to view the object's 2D or 3D contour

3. **Model** to view the object's 3D wire-frame

4. **Plan** to view the object's 2D schematic depiction while working in Plan view

5. **Schematic** to view the object as a 2D symbol

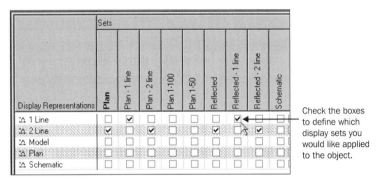

Check the boxes to define which display sets you would like applied to the object.

The following illustration shows the same object being viewed in different display representations.

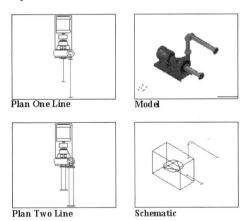

Plan One Line          Model

Plan Two Line          Schematic

Display representations are gathered together in *display sets*. A display set is a group of similar display representations for different objects. For example, if you need to plot a single line-drawing plan, you would want to use the following display representations:

- The **1 Line** display representation of the **pump** or **MvPart**.
- The **1 Line** display representation of the **Pipe**.
- The **General** display representation of the **annotation label**.

However, the display set Plan 1 line has already gathered together all of these display representations into one place.

Display sets are then assigned to *display configurations*. A display configuration assigns a display set to a view direction. The configuration is attached to a viewport. When the view direction of that viewport is set to TOP view, then the display configuration uses the display set Plan 1-line. If the view direction of the viewport changes to an isometric view, then the display configuration knows this and changes the display set to Model, and the model representation of the object is displayed on the screen.

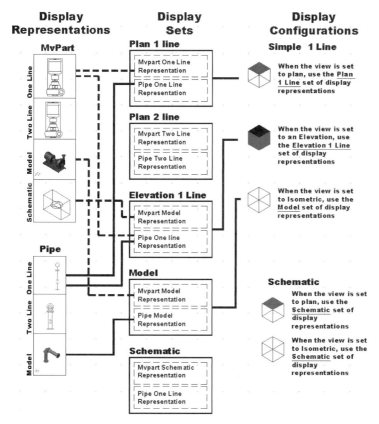

For a complete description of all the different display sets and what views they are useful for, refer to the "Building Systems Displays" section in the Autodesk® Building Systems online Help.

## AEC OBJECTS AND STYLES

The majority of AEC objects are style-based. For example, a fan contains styles. Let's say that you choose a Propeller Fan MvPart. The Propeller Fan contains several different style properties, such as size, name, connection height, connection width, and much more. Therefore, styles are a storage place for the variables associated with each of the different AEC objects.

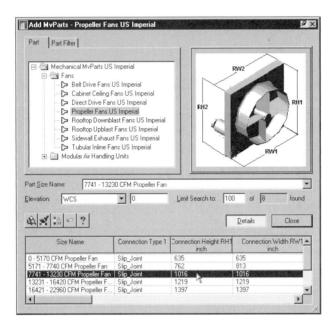

There are two general types of content that are provided for you to use with Building Systems – Content held as a style in another drawing and Content that is stored in a catalog, that creates the style in the working drawing as it is added. Both of these types of content create styles when you add them to the drawing. The differences between them are how the content is stored, and how you create new styles of these objects.

Some content is stored as a style in another drawing. Examples of these types of objects are: Schematic Symbols, Schematic Lines, Labels, and Schedule Tables. Building Systems reads these other drawings and imports the styles as you need them into your current drawing. Drawing stored styles can be created in the working drawing from scratch using the Style Manager.

Catalog based objects are stored in a catalog of parts. MvParts, Ducts and Duct Fittings, Pipes and Pipe Fittings are three different catalogs that are installed with the software. When you add an MvPart to the drawing, an MvPart style is created in the drawing based on the settings held in the MvPart catalog. You cannot create new MvPart styles directly in the drawing. You must use the Content Builder to add new parts to the catalog, and then add them to your drawing from the catalog.

## MODIFYING AN OBJECT'S STYLE

All Building Systems object styles are accessed through the Style dialog box. Depending on the Building Systems object that you select, the tabs that appear on the Style dialog box differ. For this example, an MvPart is used because MvParts contain *most* of the tabs that you will see on other Building Systems objects Style dialog boxes.

You can access the MvPart Style dialog box by selecting an MvPart, right-clicking, and clicking on Edit MvPart Style.

The General tab reflects the name and type of MvPart, and it also enables you to add extra descriptions, for example, the contact information for the manufacturer. The other important facet of this dialog box is the Property Sets button. The Property Sets button enables you to define the property set information that you would like to use when you create a schedule of the MvParts in your drawing.

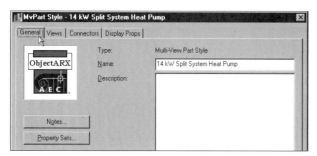

The Views tab enables you to add and remove view definitions of the MvPart. You have the ability to add view directions because MvParts are (as they are named) multi-view parts. MvParts are comprised of many different blocks. Each of these blocks is assigned a view direction. This enables the MvPart to display a different block based on the direction from which it is viewed. For example, as you change the view from top to front, the MvPart will display a different view block that represents the proper view of the part. You use the view block field to assign which way you would like a particular block (contained within your MvPart) to display using display representations. You will be using this tab frequently throughout the lessons in this book.

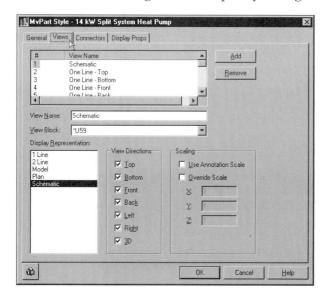

The Connectors tab enables you to add and remove domains for the connectors that are built-in to the MvPart, for example, a pipe domain. You can also add a description of that domain. The connection point information reflects where the connection is located on the object. This functionality is explained in more detail in the section, "How Building Systems Objects Connect".

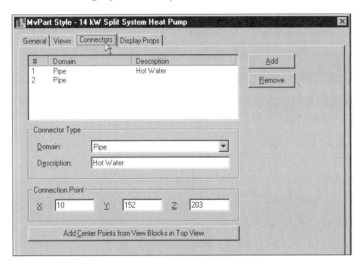

The Display Props tab enables you to customize the appearance of the MvPart in your drawing.

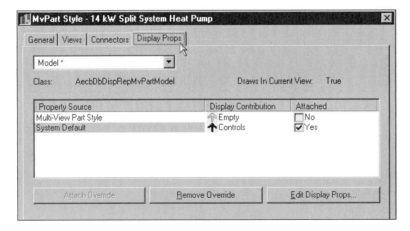

When you click the Edit Display Props button, you access the Entity Properties dialog box, which works similarly to AutoCAD's Layer Manager. You can change several subcomponents of the MvPart using the Entity Properties dialog box, including turning the display on or off, and modifying the layer name, color, linetype, lineweight, and scale factor.

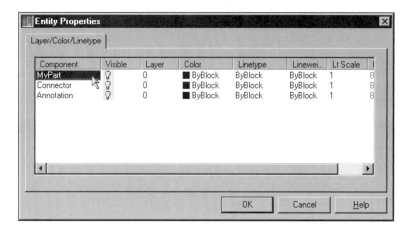

The Entity Properties dialog box gives you access to the subobjects within the MvPart, similar to changing the properties of a nested block. Although you are modifying layer, color, lineweight, linetype, and ltscale factor here, you are only changing these for the subcomponents of the MvPart. The actual layer that the MvPart is on within the drawing does not change.

When you change values using the Entity Properties dialog box, you are changing these for a specific **Display Representation**. You may have to change the same value for other display representations if your plotted sheets use other representations.

## HOW AEC OBJECTS DIFFER FROM EACH OTHER

Although Architectural Desktop objects and Building Systems objects are both called AEC objects, there are some differences between them. You may want to note which product provides what type of objects. There is little to no overlap in the type of objects provided with Architectural Desktop and Building Systems.

Architectural Desktop provides walls, doors, windows, column and ceiling grids, scheduling, and area evaluations. These are all the elements that an architect needs to illustrate a building. Building Systems adds ducts, pipes, and equipment, the parts the mechanical office needs to illustrate the mechanical design. However, Building Systems relies on Architectural Desktop to give the mechanical team access to ceiling grids, scheduling, and area evaluations. For ease of use, grids have been added to the MEP Common menu.

**Note:** On the user interface, everything to the left of the MEP Common pulldown menu belongs to AutoCAD. All Building Systems mechanical commands are accessed from the MEP Common and the Mechanical pulldown menu. The Electrical and Plumbing pulldown menus are also Building Systems menus, but will not be covered in this book. The Documentation and Desktop pulldown menus are Architectural Desktop menus that provide access to tools and functions such as schedules, sections that are common to both Achitectural Desktop and Building Systems.

## DIFFERENCES IN STORING AND ACCESSING AEC OBJECTS

Many of the Building Systems objects that are provided with the product are stored in catalogs. New styles of objects are created in the current drawing when they are needed by reading the catalog for the definition of the style. You access the catalogs using the Building Systems Catalogs tab on the AutoCAD Options dialog box. To find this tab, from the Tools menu, click Options, and then click the Building Systems Catalogs tab.

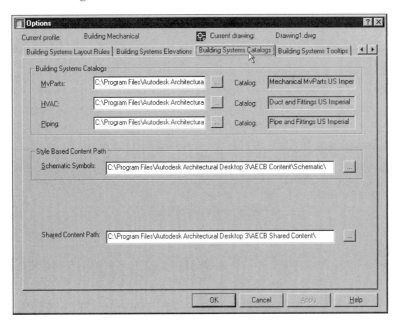

To view the different type of objects contained in a particular catalog, click the ellipses buttons to access the Select Catalog File dialog box.

The path names point to the catalog that will be used by the software. This path may be set to a networked location by your CAD manager to assure that everyone in the office is utilizing the same set of parts as they create their drawings. You probably do not want to change this setting without checking with your CAD manager or CAD lead for the project first.

All of the Architectural Desktop objects provided are stored either in the template or in content .dwg files. Architectural Desktop object style definitions (unlike most Building Systems objects) are either created by the user or imported from the sample content drawings using Style Manager. Style Manager has two panels; on the left panel you can browse to the style of the Architectural Desktop object that you would like to use, while the right panel displays information specific to that object. Style Manager only reflects Architectural Desktop object styles that are contained within a drawing that you have open. To see what the Style Manager looks like when using an architectural drawing, you can open one of the tutorial drawings that are included with Architectural Desktop by using the Help menu. Once you have opened an architectural drawing, from the Desktop menu click Style Manager.

The following illustration shows Style Manager displaying all of the door styles contained in an opened architectural drawing.

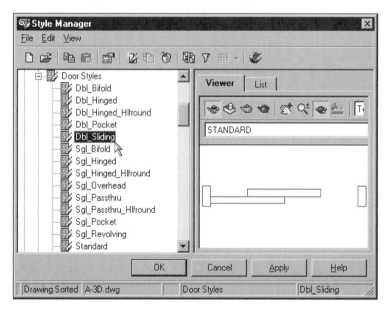

To access both Architectural Desktop objects and Building Systems objects you use the Add commands from the various pulldown menus. For example, to access an Architectural Desktop door object you would type DoorAdd at the command line.

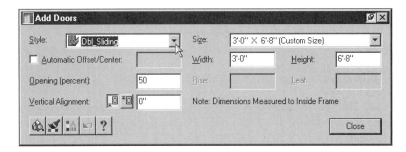

You can add any door style that you have created for your drawing from the Style pulldown menu.

To access a Building Systems object such as a duct, you would click Mechanical ➤ Duct ➤ Add.

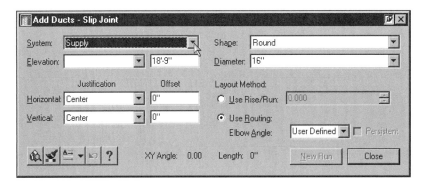

Notice that where the pulldown menu was labeled Style in the Architectural Desktop Add Door dialog box, the Building Systems Add Duct dialog box has a System pulldown instead. This is another major difference between Architectural Desktop objects and Building Systems objects.

### SYSTEMS VERSUS STYLES

Building Systems *systems* provide a way to group your Building Systems objects the way that Architectural Desktop *styles* group Architectural Desktop objects. You learn about the use of systems in conjunction with Building Systems objects in the section, "Understanding Systems".

You learned how styles are used in conjunction with Building Systems objects in the section, "AEC objects and Styles". If you want to learn how to create styles in Architectural Desktop, refer to the Architectural Desktop online Help and tutorials. However, you may never need to do this, especially if you normally work from pre-created architectural drawings. The important fact to remember is that Building Systems objects have associated styles, created by you through the catalogs instead of the Style Manager, which is how it is done in Architectural Desktop.

## ATTACHING AEC OBJECTS

Another major difference between Architectural Desktop objects and Building Systems objects are the way that the objects 'attach' to one another. In Architectural Desktop, objects attach themselves by anchoring to one another. However, since the objects know what they are, they know what they can or cannot anchor to, relieving the user from having to do this manually. The anchoring happens automatically when you add objects to a drawing. For example, if walls are drawn and you add a door, the door will automatically anchor itself to the wall.

Although Building Systems objects also know what they are, they do not become attached to each other through automatic anchoring, but through connectors that are built into all Building Systems objects. Architectural Desktop objects do not contain connectors. You will learn about the use of connectors in conjunction with Building Systems objects in the section, "How Building Systems Objects Connect".

# BUILDING SYSTEMS OBJECTS

Building Systems objects are system-based objects.

All Building Systems objects contain:

- **Mechanical industry** specific equipment, fittings, and accessories are the basis of all Building Systems objects. You need to use Architectural Desktop objects such as ceiling grids and spaces if you are not working off of a pre-created architectural drawing.

- **Connectors** are built into all Building Systems objects and provide not only the ability to connect the parts in your drawing, but also verify that you are connecting the proper parts. You will learn more about connectors in the section, "How Building Systems Objects Connect".

- **Labeling** commands included in Building Systems have the ability to read the properties that you assigned to any Building Systems object. In addition to the labeling features in Building Systems, you can use Autodesk® Architectural Desktop annotation features such as title marks and leader lines.

- **Scheduling** your drawing can be accomplished using any Building Systems object; however, the amount of information that the schedule can contain depends on what type of object you are working with (schematic or MvPart).

## SCHEMATIC SYMBOL AND LINE OBJECTS

Schematic symbols and lines are used in the creation of schematic diagrams. Schematic diagrams are single-line, 2D drawings. Unlike MvParts, where you can use the compass feature, you use the Schematic Symbol dialog box to place symbols at a specific rotation angle. Schematic lines and schematic symbols always maintain some sort of anchoring relationship to each other. It just depends on the order in which you place the lines and symbols into the drawing.

The following illustration shows the heating water system schematic that you will be drafting in Chapter 3.

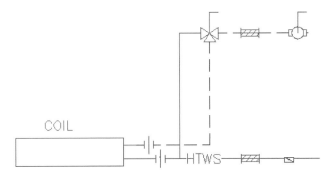

Schematic lines and symbols are assigned to different systems, and assigned identifications (IDs) for annotation purposes.

Schematic symbols have a limited amount of information that can be scheduled. Also, the connectors that are built into schematic symbols and lines are only used to help place lines and symbols in relationship to each other. They do not pass information from part to part as connectors on Ducts, fittings and MvParts do.

You will learn more about schematic lines and schematic symbols in Chapter 3, "Creating a Mechanical Schematic Diagram".

## MVPART OBJECTS

MvParts (multi-view parts) represent the equipment and accessories (such as fans, coils, pumps, and gauges) in the design. MvParts are a collection of AutoCAD blocks. Each side of an MvPart object has a different block assigned, so that you can see the object from any direction. In addition to the AutoCAD blocks, an MvPart has connectors. The connectors on an MvPart contain specific domain information, and therefore each connector knows whether or not it should be attached to a duct or pipe. The connectors also have the ability to recognize and pass on size and system information to the objects that they connect to. For example, a 6"x 6" base-mounted pump contains the size (6") defined for the inlet and outflow of the pump within the connectors.

When MvParts are first placed in your drawing, the system of the connectors are "undefined". This allows you to change the system of the part after you place it in the drawing. You do not have to specify the system of the MvPart, you can just place the MvPart in the drawing. The Undefined system is a system that enables you to attach any duct or pipe of any system to the MvPart.

MvParts also have information associated with them called Catalog Entries. The catalog entry information is what differentiates the particular part from other parts in the catalog. You can see the list of catalog entries in the following illustration.

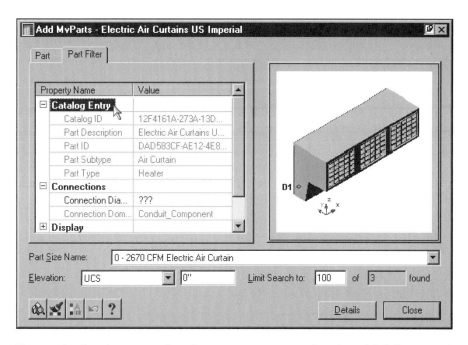

Due to the fact that so much information is associated with each MvPart, you have the ability to schedule almost anything that you want. You also have the ability to assign systems to MvParts that help to define what all of the different mechanical drawing elements belong to. For example, a coil MvPart may belong to a heating water system. You will learn more about systems in the section, "Understanding Systems".

### DUCT AND PIPE OBJECTS

Duct and pipe objects are used in conjunction with MvParts. They have similar qualities to MvParts in that they have multiple display representations; however, in the case of a duct or pipe, there are only four representations—1-Line, 2-Line, Model, or Haloed line. Ducts and pipes are assigned a system as you draw them. While MvParts are catalog-driven, what is placed in your drawing is a collection of AutoCAD blocks that are a finite set of MvParts that you may choose from the catalog. When you draw a duct or pipe your drawing is dynamic. The duct or pipe reads the size from the add dialog box and gets its length from the start and end points that you pick on the screen. Ducts and pipes are also catalog-driven, but you can exceed the values of the catalog. For example, you may need a duct that is 120" x 80". While the catalog does not have ducts that size, because it is dynamic, you may still draw a large duct. Ducts or pipes may be resized after they are drawn, and the size will get passed along the connected ducts and pipes in the branch if you tell it to do so. Ducts and pipe also have connectors at each end.

## FITTING OBJECTS

As you draw ducts and pipes, fittings will be placed at each turn you create. The type of fitting is determined in the Duct Add Preferences Dialog box. There will be times when you want to change the type of fitting as you are adding ductwork. For example, you might want to place a smooth radius elbow in one location, and a mitered elbow in another. You may change the default fitting at any time, or you may place them manually in the drawing.

The left side of the following illustration shows a Rectangular Duct Slip Joint segment *before* being connected to any other duct segments. The right side of the illustration highlights the 90 Degree Rectangular Mitered Elbow Slip Joint that was automatically added when the duct run direction changed by 90 degrees. The illustration also displays the building system snaps that were used to connect the duct to the fitting.

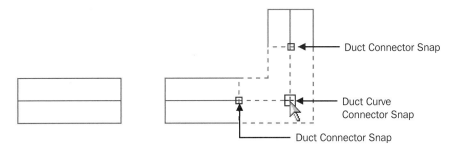

When you draft duct and pipe runs you use the compass feature to guide the direction and angle of your run. The compass also enables you to place MvParts equipment at specific rotation angles in your drawing. You will be using the compass more in Chapter 4, "Creating a Mechanical Plan".

## HOW BUILDING SYSTEMS OBJECTS CONNECT

All Building Systems objects have connectors. These connectors are built into the object, and by using them appropriately, are what make each piece of your mechanical layout know what it is and how it relates to your entire mechanical design. Without the connectors, your building design is not a complete network. If Building Systems objects did not have connectors, each piece of your design would stand-alone; the duct would not know that it was connected to an air-handling unit, and the air-handling unit would not know its purpose.

The five main purposes of connectors are:

1. To control whether a duct or a pipe is connected to an MvPart.
2. To check to see if the proper connection type is set for the pipe or duct. For example, a pipe connection could be flanged, threaded, or brazed (to name a few), and a duct connection could be banded, slip joint, or vanstone.

3.  To pass the proper size information on to the Add Pipe or Add Duct dialog box. This automatically fills in the information in the Add dialog box if the duct or pipe is drawn from an MvPart.

4.  To specify and add the appropriate transition in the case where you are ending a pipe or duct run on an MvPart.

5.  To locate appropriate snap points on objects. Connectors work in conjunction with building systems snaps when adding pipes and duct.

Each Building Systems object has a different amount of connectors, depending on the need of the object. For example, a ceiling terminal only has one connection, but a 3-way valve has three.

The following illustration shows a 6" flanged 3-way valve with its associated connection points.

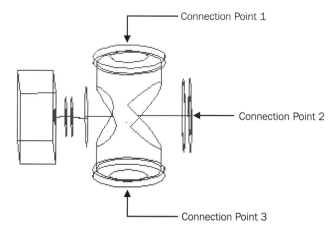

Each of the connectors on an object has a predefined domain. For example, all three of the connection points on the 6" flanged 3-way valve have a pipe domain specified. Therefore, you cannot add a duct to any of these connectors. However, an air-handling unit coil module has four connection points, two predefined with duct domains, and two predefined with pipe domains.

All connection points that are on duct segments, pipe segments, and fittings are located at the start and end of the object.

Building systems snaps work in conjunction with the connection points on the objects. Building systems snaps have the ability to recognize where the connection points are on Building Systems objects. It is important that you use building systems snaps when you draft your mechanical layout. For example, you *could* use AutoCAD's® endpoint object snap to connect the end of the duct to the duct fitting. By default, all of the building systems snaps are already turned on for you. You can access the different types of building systems snaps by clicking the Tools menu and then clicking

Drafting Settings. Click on the Building Systems Snaps tab on the Drafting Settings dialog box to view the building systems snaps that are available to you. It is also helpful to familiarize yourself with the snap glyph that displays when you are connecting the different parts of your drawing.

The glyph that displays when snapping to the curve of a duct.

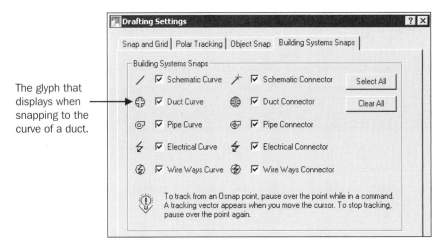

## UNDERSTANDING SYSTEMS

There are many different mechanical elements contained in a building. Building Systems enables you to define the different mechanical elements through the use of systems. For example, one mechanical drawing could contain a hot water system, a cold water system, an air supply system, and an air return system.

Although you do not need to, it is easier to choose the systems that you want to use in your drawing at the start of a project. There are predefined HVAC and Piping systems in Building Systems that are based on industry standards, for example, SMACNA. The systems that are included in Building Systems contain system abbreviations, layer keys, and display properties. In order to access these predefined HVAC systems you need to be using a Building Systems template, such as the *Aecb Building Model (Imperial)3.dwt* template. You can start a new drawing using this template by selecting the *Aecb Building Model (Imperial)3.dwt* template in AutoCAD's Today dialog box.

**Note:** There are no predefined systems in the AECB Schematic template. The *Aecb Building Model (Imperial)3.dwt* contains sample styles for all three disciplines, Mechanical, Electrical and Plumbing.

System definitions are accessed through the Style Manager. You can access the Style Manager on the Mechanical menu by clicking Mechanical Systems ➤ HVAC System Definitions or Pipe System Definitions.

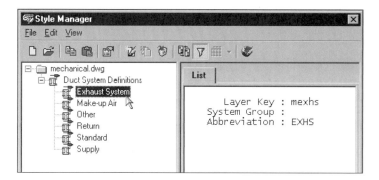

The quickest way to create your own system is to base it on the predefined system that is closest to your needs. You do this by selecting the system in Style Manager, right-clicking and clicking Copy, and then right-clicking and clicking Paste. Copying a predefined system style saves you time because some of the information that you need for your new system will already be defined for you. You can then change any necessary information by selecting the copied system, right-clicking and clicking Edit.

In the System Definition dialog box you can edit the system by changing information on the following tabs:

- **General tab** – reflects the type of system, the name of the system, and also enables you to add extra descriptions; for example, the shape of the ductwork that you want used.

- **Design Rules tab** – enables you to define a system abbreviation that will be used when you label your drawing. This tab is also used to create system groups, and to assign the layer key style that you want used.

- **Rise and Drop tab** – enables you to see the type of system that the duct or pipe is associated with when viewing your drawing in Plan view (looking down on a 'flat' drawing). For example, if you are working with a supply system, then you will see an **X** displayed at the top of your duct. If you were working with an exhaust system, then you would see a **Y**.

- **Display Props tab** – enables you to customize the appearance of the system in your drawing. When you click the Edit Display Props button, you access the Entity Properties dialog box, which works similarly to AutoCAD's Layer Manager. You can change the layer, color, and linetype of the system using the Entity Properties dialog box.

Once you have defined the systems that you want to use, you 'assign' each Building Systems object to a system definition as you add it to the drawing. For example, when you add a piece of duct, you select the system that you would like to associate it with in the Add Ducts dialog box.

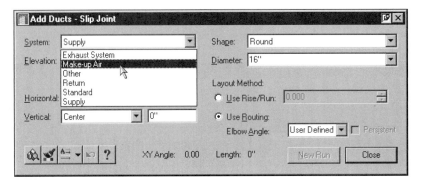

# KEY CONCEPTS: UNDERSTANDING AEC OBJECTS

AEC objects are specific to the architectural, engineering, and construction industries. They are not only symbolic representations of an object; they *are* the object.

You can display the same AEC object differently in different views.

AEC objects respect and relate to one another, and you can use and create different styles of the same AEC object to customize your work environment.

AutoCAD entities are drawing elements that are not domain-specific, and do not have the ability to recognize what they are.

AEC objects contain display representations. Depending on the display needs of the object, the object can have one or several display representations associated with it.

Display representations are managed through Architectural Desktop's Display Manager.

Display representations are gathered together in *display sets*.

A display set is a group of similar display representations for different objects.

Display sets are then assigned to *display configurations*.

A display configuration assigns a display set to a view direction, and then the display configuration is assigned to a *viewport*.

The majority of AEC objects are style-based. Styles are a storage place for the variables associated with each of the different AEC objects.

Although Architectural Desktop objects and Building Systems objects are both called AEC objects, there are some differences between them, such as where in the product they are stored and how they attach to each other.

All Building Systems objects have connectors that are built into the object.

Connectors are what make each piece of your mechanical layout know what it is, and how it relates to your entire mechanical design.

Each of the connectors on an object has a predefined domain, such as duct or pipe.

Building systems snaps work in conjunction with the connection points on the objects and have the ability to recognize where the connection points are on Building Systems objects.

Building Systems enables you to define the different mechanical elements through the use of system definitions.

There are predefined system definitions in Building Systems templates that are based on industry standards.

MvParts are inserted in the drawing with their connectors assigned to the "undefined" system. You may leave them as is or change them after you place them in the drawing.

You assign each duct and pipe to a system as they are added to the drawing.

# CHAPTER 3

# Creating a Mechanical Schematic Diagram

After you read this chapter, you should understand the following:

- how to specify the default options for your mechanical schematic drawing
- how to draft a mechanical schematic drawing by inserting schematic lines and symbols
- how to modify a mechanical schematic drawing by changing the properties of a schematic symbol, and by using AutoCAD commands such as Move, Copy, Trim, and Extend
- how to create schematic line and symbol styles in order to customize your drawing
- how to annotate the schematic drawing using the Label commands in Building Systems, as well as Architectural Desktop Title and Leader commands

## WHAT IS A SCHEMATIC DIAGRAM IN BUILDING SYSTEMS?

Building Systems provides a set of tools to help you produce schematic diagrams. Unlike other aspects of Building Systems, the schematic tools produce flat 2D drawings. Use the schematic tools in Building Systems to prepare water circulation diagrams, exhaust air diagrams, and other schematic or diagrammatic drawings. The Schematic tools can also be used as a preliminary one-line design sketch tool. The schematic tools provided with Building Systems are schematic symbols and schematic lines. The schematic lines and symbols aid in quickly producing single-line preliminary duct layouts during the early design phase of the project.

Autodesk® Building Systems differs from AutoCAD in that it provides you with symbols to represent pieces of schematic HVAC equipment, fittings, and controls. These pieces of schematic HVAC equipment, fittings, and controls are categorized and called schematic symbols. These symbols are connected with schematic lines that represent associated ducts and pipes in the schematic drawing.

## UNDERSTANDING SCHEMATIC SYMBOLS AND SCHEMATIC LINES

Schematic symbols are Building Systems objects. They are style-based and system-governed. What you see displayed on the screen when you place a schematic symbol is an AutoCAD block that is part of the style definition. One or more connector points that aid in drawing the schematic diagram are also contained in the schematic symbol style. A special set of OSNAPs utilizes these connectors, enabling you to draw the layout more easily. The connectors are also the basis of the anchors that operate between the schematic lines and the schematic symbols.

You can place symbols at a specific rotation angle, or you can specify the rotation on the screen. The schematic symbol will then mask the segment of the schematic line that the schematic symbol has been anchored to, giving the appearance that the schematic line has been broken to make way for the schematic symbol. When the schematic symbol is erased from a schematic flow diagram that has been constructed in Building Systems, the schematic line will appear as it did prior to inserting the schematic symbol. You have the ability to assign properties to a selected symbol, such as a system label and identification (ID). The ability to assign these properties is important when differentiating the mechanical systems contained in the drawing. For example, you can assign a coil symbol a system label of HTW for Heating Water System. The system label and ID are used when annotating your drawing with Building Systems Labels.

A schematic line has the look and feel of a basic AutoCAD® polyline. However, there are several differences between a schematic line drawn using Building Systems, and a basic AutoCAD line. Schematic lines drawn using Building Systems have the ability to recognize connection points that are included on schematic symbols by using building systems snaps, such as the connector snap. Like schematic symbols, you also have the ability to assign properties to schematic lines, such as a system label and designation ID. Schematic lines also have the ability to automatically show crossed lines as breaks. The connection points that are built into schematic lines are called schematic curve connectors, and are used when placing schematic symbols onto schematic lines.

## Schematic Systems

Both schematic lines and schematic symbols are style-based AEC objects. These styles contain many variables about how these objects behave in the drawing. As schematic lines and symbols are placed in the drawing, they are also given a schematic system definition. The schematic system defines the layer the schematic line or symbols is placed on. The schematic system holds a system abbreviation that can be used to label the line or symbol. The system can also be used to control the linetype of the schematic lines.

## Orthogonal and Isometric Schematics

When drawing lines, you can select the Isometric icon, and use the Isometric control in the Add Schematic Line dialog box to automatically draw your lines at 30 degrees to horizontal.

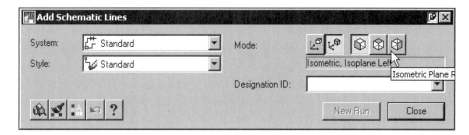

The mode right, left, or top is stored in the line. When you add a schematic symbol that is Isometric-enabled it will read its orientation from the line and orient itself to the right, left, or top mode stored on the line. Only a symbol created to use this feature will have the Isometric symbol available within the Add Schematic Symbol dialog box. The other symbols will have the Isometric icon grayed out. Those symbols that have the Isometric enabled start with 3-way valves.

The Labeling commands included in Building Systems have the ability to read the properties that you assigned to the lines and symbols. You can label symbols using the system label, ID, and style name. You can label lines using System Label from the Schematic System Definition, Style Name, and Designation from the Schematic Line Style. In addition to the labeling features in Building Systems, you can use Autodesk® Architectural Desktop annotation features such as title marks and leader lines.

Building Systems works in conjunction with Architectural Desktop to enable you to create your own symbol and line styles using Architectural Desktop's Style Manager. This feature is important if the symbol necessary for your drawing is not included in the schematic symbol part catalog.

Lines and symbols are anchored to each other; whether the symbol is anchored to the line or the line is anchored to the symbol depends on the order in which you place them in your drawing.

A SCHEMATIC SYMBOL PLACED USING THE LINE'S SCHEMATIC CONNECTOR IS ANCHORED TO THE LINE.

A SCHEMATIC LINE PLACED USING A SCHEMATIC SYMBOLS CONNECTOR IS ANCHORED TO THE SYMBOL.

If you place the symbol onto a pre-drawn line, then the symbol is anchored to the line. The symbols will mask the line, hiding the line behind it. A symbol that is anchored to a line can be placed anywhere on a line. This means that if you move the line, the anchored symbol moves with the line. If you erase the line, then the symbol is also erased.

If you place a symbol away from any lines in the drawing, then the symbol does not become anchored to the line when it is drawn. Therefore, if you erase the line the symbol is not also erased.

To summarize, a symbol that is placed in a drawing *before* a line is *not* anchored to a line. A symbol that is placed *after* a line is drawn *is* anchored to the line to which it is placed.

The reverse is true when placing lines; a line drawn *before* a symbol is placed is *not* anchored to the symbol. A line that is drawn *after* a symbol is placed *is* anchored to the symbol.

Therefore, lines and symbols always maintain some sort of anchoring relationship to each other. It just depends on the order in which you place the lines and symbols into the drawing.

**LESSON OBJECTIVES**

This chapter contains five lessons. The lessons are broken down into one or more exercises. Most exercises have a corresponding drawing file that is located on the CD-ROM included with this book.

Lessons in this chapter include:

- **Lesson 1: Specifying your Schematic Drawing Options**

  Here you will learn how to specify your snap settings as well as assign the default options appropriate for a mechanical schematic drawing.

- **Lesson 2: Drafting your Mechanical Schematic Diagram**

  You will learn how to draft a mechanical schematic drawing that represents the piping schematic of a reheat coil for a VAV control box.

- **Lesson 3: Modifying your Mechanical Schematic Diagram**

  You will change the symbol scale and use basic AutoCAD commands such as trim and move to modify a mechanical schematic drawing.

- **Lesson 4: Creating Schematic Line and Symbol Styles**

  You will become familiar with using Style Manager to create new styles and to access the style definitions.

- **Lesson 5: Annotating your Mechanical Schematic Diagram**

  You will annotate a heating water system schematic drawing using the labeling commands included with Building Systems and Architectural Desktop's documentation symbols.

## LESSON 1: SPECIFYING YOUR SCHEMATIC DRAWING OPTIONS

Like AutoCAD and Architectural Desktop, Building Systems provides you with the option of using a template. The advantage of using a pre-defined template is that many of the desired option settings are already specified for you. Templates also provide you with layout tabs other than just the Model tab.

In this lesson you will learn to specify your snap settings, as well as assign the default options appropriate for a mechanical schematic drawing using the Building Systems–specific tabs contained on the AutoCAD Options dialog box.

### Create a Drawing from a Template

1. Access the AECB Schematic template:
   a. From the File Menu, click New.
   b. In the Today dialog box, click the Create Drawings tab under My Drawings.
   c. Under Select How to Begin, click Template and then click Browse.

2. Open the AECB Schematic template:

   a. In the Select File dialog box, select the *Aecb Schematic (Imperial) 3.dwt* template.

   b. Click Open.

   c. Verify that the drawing displays three layout tabs: Model, Schematic, and Plot Schematic.

**Note:** We recommend that you draft your mechanical schematic drawing using the schematic layout tab. You can then use the Plot Schematic layout tab when the drawing is ready to be plotted.

## Set the Drawing Tools Settings

3. Specify the snap settings:

   a. At the bottom of the screen, right-click on the word SNAP, and click Settings.

   b. On the Drafting Settings dialog box, click the Building Systems Snaps tab.

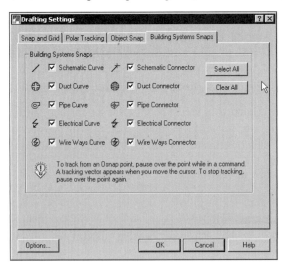

   c. Verify that all of the building systems snaps are selected and then click OK.

4.  Turn off Object Snaps (OSNAP):

    Turning off OSNAP ensures that you are snapping to the connectors that are built into the symbols.

    a.  At the bottom of the screen, hover your cursor over the word OSNAP to display the Object Snap tooltip.

    b.  Verify that <OSNAP OFF> displays on the command line.

5.  Specify Building Systems options for a schematic diagram:

    a.  From the Tools menu, click Options.

    b.  Click the Building Systems Catalogs tab, and verify that the Schematic Symbols Style-Based Content Path is set to *C:\Program Files\Autodesk Building Systems 3\AECB Content 3\Schematic\.*

        Click the ellipses [...] button if you would like to see the schematic drawings available. Each one of the drawings in this folder contains multiple schematic symbol styles. When you add a schematic symbol, the dialog box looks into each of these drawings and lets you select individual symbols to place in your drawing.

    c.  Click the Building Systems Tooltips tab, and select Object Class, Object, Name, and System, and then click OK.

You now know how to specify the default options for a mechanical schematic drawing. The other Building Systems tabs available to you on the AutoCAD Options dialog box are explained later in the text when applicable to a particular lesson task.

## LESSON 2: DRAFTING YOUR MECHANICAL SCHEMATIC DIAGRAM

This lesson contains three exercises, with a drawing file associated with it. When you complete the exercises in this lesson you will have drafted a mechanical schematic drawing that represents the piping schematic of a reheat coil for a VAV control box.

To accomplish this task you will be using the Add Schematic Symbols command and the Add Schematic Lines command. At the end of each exercise there is a screen shot of what your finished drawing should look like.

This lesson contains the following exercises:

-   Placing Schematic Symbols
-   Drawing Schematic Lines
-   Adding Schematic Symbols to Schematic Lines

### EXERCISE 1: PLACING SCHEMATIC SYMBOLS

In this exercise you will be inserting a coil and two flexible connection pipe fittings. During this process you will be specifying a system and ID to represent a heating water system. The system will determine the layer the symbols are placed on, along with the system abbreviation that may be used to label the symbol later. The ID is held on the symbol itself and will also be used later for labeling. The symbols that you place in this drawing are not anchored to the lines. This is because you are placing the symbols *before* you are drawing the lines.

 **Open:** *03 Placing Schematic Symbols.dwg*

### Add a Coil to the Drawing

1. Select a coil to add to your drawing:

   a. From the MEP Common Menu, click Schematics ➤ Add Schematic Symbol.

   b. Under Schematic Symbol, select HVAC Equipment.

   The selections in this drop-down list contain predefined groupings of the symbols that are provided with Building Systems. These groupings reference the Schematic Symbols Style-Based Content Path that you verified on the Building Systems Catalogs tab in lesson 1.

   c. Scroll down to select the Coil symbol.

2. Specify the add coil properties:

   The first insertion properties assigned become the default for any other symbols or lines that you place in your drawing.

   a. Verify that the Details button has been clicked to expose the Properties side of the dialog box.

   b. From the system pull-down, select Equipment for the system.

c.    For ID, enter **Coil**.

3.    Insert the coil into the drawing:

a.    Deselect Specify Rotation on Screen.

b.    For Rotation, enter **90**.

c.    Click once on the screen, and then pick a point in the lower-left corner of the screen to place the coil.

Clicking on the screen takes the cursor focus out of the dialog box and onto the AutoCAD drawing screen. This takes a bit of getting used to. Most of the commands in Building Systems will have you working in dialog boxes to add anything. If you have the thumb tack toggle at the upper-right corner of the dialog box out, then the dialog box will minimize as you move your cursor off of it, and you do not need this extra pick to get back into the drawing. In general, the steps in this book assume you are working with the thumbtack *in*.

### Add a Flex Connection Pipe Fitting with Otrack

4.    Select a flexible pipe fitting to connect to the coil:

a.    Click once in the dialog box to make it active.

b.    In the Schematic Symbol drop-down list, choose Piping Fittings.

c.    Click the Miscellaneous button and select Flex Connection Pipe Fitting.

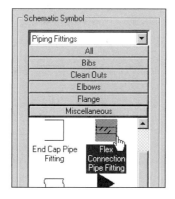

5.  Specify the add flex connection pipe fitting properties:

    Notice that the properties that you assigned to the coil remain because they became the default.

    a.  Leave the System as Equipment.

    b.  Change the ID to Flex Connection.

6.  Specify the insertion values for the flex connection pipe fitting:

    a.  Select Specify Rotation on Screen.

    b.  Click the Thumbtack at the upper-right hand corner of the dialog box.

        Clicking the thumbtack allows you to work interactively between the dialog box and the AutoCAD screen.

    c.  Change the Justification to Middle Left.

7.  Insert the flex connection pipe fitting into the drawing:

    a.  Turn on Osnap, verify that the Endpoint snap mode is selected, and then select Object Snap Tracking On. This can be accomplished by right-clicking the OSNAP button at the bottom of the screen and selecting settings.

    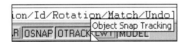

    b.  Use Object Tracking to set a temporary track point by hovering the cursor over the endpoint of the lower coil connecter, and then drag the cursor to the right and enter 18 at the command line to place the pipe fitting 1' 6" from the coil.

    c.  Rotate the pipe fitting to the right.

    d.  Click to place the pipe fitting at the correct rotation angle.

8.  Copy the flex connection pipe fitting:

    a.  Select the flex connector, right-click, and click Copy Selection.

b. Select the endpoint of the original pipe fitting as the base point and drag your cursor upwards.

c. Enter **18"** at the command line.

This is what your finished drawing should look like.

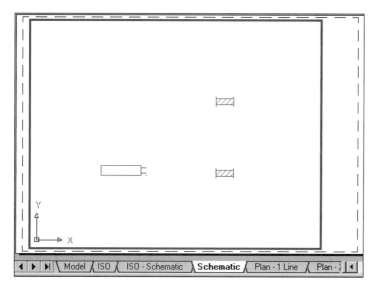

You have completed the task of inserting a coil and two flexible connection pipe fittings. In the next exercise you will be drawing lines that represent flexible piping to connect the symbols.

### EXERCISE 2: DRAWING SCHEMATIC LINES

In this exercise you will connect the coil and two flexible connection pipe fittings with lines that represent pipes. This piping serves as the heating water supply lines for your reheat coil.

**Open:** *03 Drawing Schematic Lines.dwg*

It is important to turn off OSNAPs before beginning the steps in this exercise. As long as the building system snaps are turned on in the settings, they will still be active when using a building system command, even if you have turned off OSNAPs with F8, or the toggle on the status bar.

Turn on tracking (Otrack) and turn off Polar. When you add schematic lines, you will use the Building Systems compass instead.

### Add a Schematic Line

1. Specify the schematic line properties:

   a. From the MEP Common Menu, click Schematics ➤ Add Schematic Line.

   b. Verify that the System and Style are Standard.

   c. Verify that the mode is orthogonal, and delete any information in the Designation ID field.

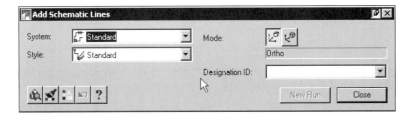

2. Draw the first schematic line from the coil:

   a. Pick the lower connection of the coil using the schematic connector.

   b. Drag the cursor to the right and pick the connector on the left side of the flex pipe symbol. Continue to drag the cursor through the flex pipe symbol and pick the connector on the right side of the pipe fitting.

   c. Continue to draw the line to the right edge of the drawing and press ENTER.

3. Open the Add Schematic Lines dialog box and verify the schematic line properties:

   a. From the MEP Common Menu, click Schematics ➤ Add Schematic Line.

   b. Verify that the System and Style are Standard.

   c. Verify that the mode is orthogonal, and that there is no information in the Designation ID field.

### Add a Schematic Line with the Compass

4. Draw the second schematic line from the coil:

   a. Pick the upper connection of the coil using the schematic connector.

   b. Drag the cursor to the right and pick a point to change the direction of the line so that the line can move upwards towards the top pipe fitting. Use the compass and Otrack to help guide the direction of your run.

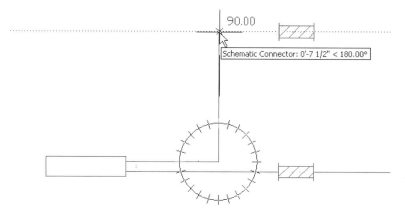

c. Drag the cursor to the right and pick the left side of the pipe fitting. Continue to drag the cursor through the pipe fitting and pick the right side of the pipe fitting.

d. Continue to draw the line to the right edge of the drawing and then press ENTER.

This is what your finished drawing should look like.

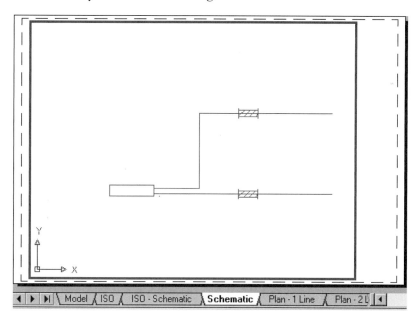

In this exercise, you connected the coil and two flexible connection pipe fittings with lines that represented piping. This piping serves as the heating water supply lines for your reheat coil. In the next exercise you will be placing more symbols into your drawing. However, this time you will be placing the symbols on to the lines.

### *EXERCISE 3: ADDING SCHEMATIC SYMBOLS TO SCHEMATIC LINES*

In this exercise you will be adding a three-way valve and a union pipe fitting to the schematic heating water system. You will anchor the symbol to the line as you insert the symbol into your drawing. This means that if you move the line, the anchored symbol moves with the line. If you erase the line then the symbol is erased as well.

**Open:** *03 Adding Symbols to Schematic Lines.dwg*

### Add a 3-Way Lever to the Schematic Line

1.  Select a three-way valve to add to your drawing:

    a.  From the MEP Common Menu, click Schematics ➤ Add Schematic Symbol.

    b.  Under Schematic Symbol, select 3 way valves.

    c.  Select the 3-Way Lever.

2.  Specify the three-way lever properties:

    a.  Verify that the Details button has been clicked to expose the Properties side of the dialog box.

    b.  Verify that the mode is ortho. This is one of the symbols that can also be used with isometric lines, but we are drawing an orthogonal diagram here.

    c.  For System, use the drop-down to select Equipment if it is not already selected.

    d.  For ID, enter **21**.

3.  Insert the three-way lever into the drawing:

    a.  For Justification, select insertion point.

    b.  Verify that Specify Rotation on Screen is deselected and that the Rotation is set to 0.

    c.  Click once on the screen. Notice that the building systems snaps appear as you hover the symbol over the line. Pick any point to the right of the pipe fitting on the top line.

    d.  Pick close.

    The three-way lever fitting is now anchored to the line.

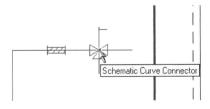

4.   Relocate the three-way lever to the left of the pipe fitting:

   a.   Select the three-way lever, right-click, and click Move.

   b.   Drag the cursor to the left of the pipe fitting. Notice that even if you pull the symbol away from the line that it still snaps to the line when you pick your point.

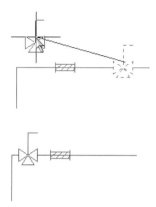

The symbol is now placed on the line at the appropriate location. The reason you were able to move the symbol so far away from the line and still have it snap to the line is because the symbol is anchored to the line. The anchor was established when the symbol was placed using the building system Schematic Curve snap. The anchor is the relationship between the line and the symbol, and is responsible for allowing the symbol to mask the line, even when it is moved along the line. If you had the AutoCAD end-point or near snap activated, the anchor would not have been created. The anchor will let you move symbols along any line that has the same System Definition, but will not let you move the symbol to a line with a different system definition.

**Add a Union Fitting Symbol**

5.   Select a screwed union pipe fitting to add to your drawing:

   a.   From the MEP Common Menu, click Schematics ➤ Add Schematic Symbol.

   b.   Under Schematic Symbol, select Piping Fittings.

   c.   Click Unions and select the symbol Screwed Union Pipe Fitting.

6.   Verify the screwed union pipe fitting properties:

   a.   Verify that the Details button has been clicked to expose the Properties side of the dialog box.

   b.   Verify that the system is set to **Equipment**.

   c.   Verify that **21** is defined for ID.

7. Insert the screwed union pipe fitting into the drawing:

   a. Verify that Justification is set to insertion point.

   b. Verify that Specify Rotation on Screen is deselected and that the Rotation is set to 0.

   c. Click once on the screen. Notice that the building systems snaps appear as you hover the symbol over the line. Pick a point to the right of the bottom coil connector.

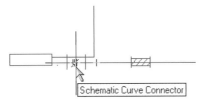

The screwed union pipe fitting is now anchored to the line. This is what your finished drawing should look like.

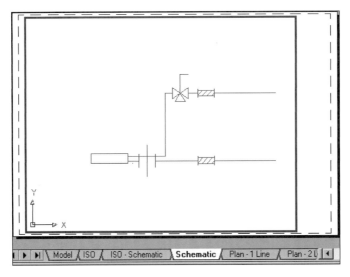

In Lesson 2, "Drafting your Mechanical Schematic Diagram," you completed three exercises: Placing Schematic Symbols, Drawing Schematic Lines, and Adding Schematic Symbols to Schematic Lines. You have now started drafting a mechanical schematic drawing that represents the piping schematic for a reheat coil for a VAV control box. There are some problems with this diagram, though. The Union is way out of proportion with the other symbols, and we need to create a bypass route through the 3-way valve. In the next lesson you will be modifying the mechanical schematic drawing.

## LESSON 3: MODIFYING YOUR MECHANICAL SCHEMATIC DIAGRAM

This lesson contains two exercises, each of which has a drawing file associated with it. When you complete the exercises in this lesson you will have modified the mechanical schematic drawing that you drafted in lesson 2, which represented the piping schematic for a reheat coil for a VAV control box.

To accomplish this task you will be using the Modify Schematic Symbols command and the Modify Schematic Lines command. You will also be changing the symbol scale, and using basic AutoCAD commands such as Trim and Move. At the end of each exercise there is a screen shot of what your finished drawing should look like.

This lesson contains the following exercises:

- Modifying Schematic Symbol Properties and Scale
- Using AutoCAD Commands to Modify Schematic Symbols

### EXERCISE 1: MODIFYING SCHEMATIC SYMBOL PROPERTIES AND SCALE

In this exercise you will be changing the system label and ID of the screwed union pipe fitting and adding a description to the flexible connection pipe fittings. During this process you will be using the Modify Schematic Symbols dialog box, as well as the Schematic Symbol Properties dialog box.

**Open:** *03 Modifying Schematic Symbols.dwg*

**Modify the Properties of the Schematic Symbols**

1. Change the system label and ID of the screwed union pipe fitting:

   a. From the MEP Common Menu, click Schematics ➤ Modify Schematic Symbol and select the union pipe fitting.

   b. Change the ID to **22**.

   c. Click Apply and then click OK.

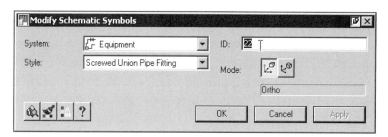

2. Modify the properties of the flexible connection pipe fittings:

    a. Select one of the flexible connection pipe fittings, right-click, and click Schematic Symbol Properties. Click the General tab and enter **3/4" Threaded copper flex**.

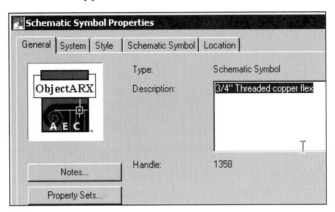

    b. Click the Style tab to view a list of available symbol styles. You can change the symbol that displays in the drawing here.

**Caution:** Although you can change the symbol in your drawing by using this tab, the results are not predictable. This is due to the fact that each symbol contains different connectors, so the new symbol that you choose for your drawing may not have the same amount of connectors or the location of the connectors may be different.

    c. Click the Locations tab and view the X, Y, and Z insertion point values, as well as the rotation angle of the flex connection pipe fitting, and then click OK.

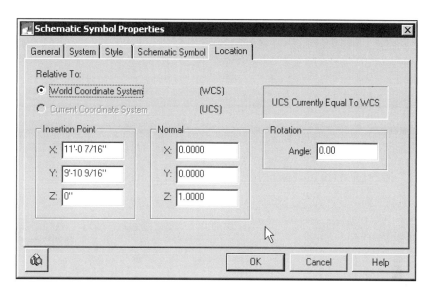

The Location tab on this dialog box is the same as the Location tab available in Architectural Desktop, and is only available after a part has been added to a drawing. If you need to adjust the rotation of the symbol, you can enter the value here.

3.   View the three-way valve properties:

a.   Select the three-way valve, right-click, and click Schematic Symbol Properties.

b.   Notice that the General, Style, and Schematic Symbol tabs contain the same options as the flex connection pipe fitting.

c.   Click the Anchor tab, view the available options, and click OK.

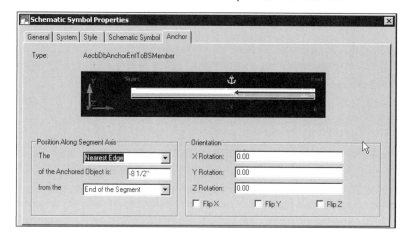

Schematic symbols that are added to a line become anchored to the line. That is why there is an Anchor tab rather then a Location tab. Since the symbol is anchored to the line, you can only change the positioning and rotation of the anchored symbol along the line. However, changing any of the values on the Anchor tab does not affect any of the location values of the line itself.

**Change the Scale of a Symbol Style**

4. Change the scale of the screwed union pipe fitting:

   a. Select the screwed union pipe fitting, right-click, and click Edit Schematic Symbol Style.

   b. Click the Views tab, and deselect Use Annotation Scale.

   c. Select OK.

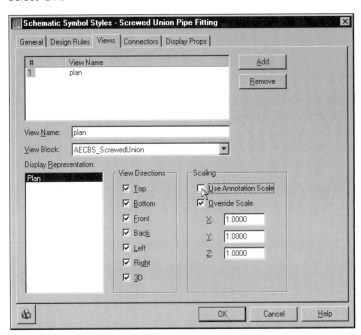

   d. Notice that the screwed union pipe fitting becomes much smaller.

   The schematic symbol scale is set in the style. By default, symbols use the drawing and annotation scale specified on Architectural Desktop's Drawing Setup dialog box. This way, if you change the drawing setup scale, then all of the symbols in the drawing are also scaled to match. When the Use Annotation Scale check box is deselected, the dynamic scaling for the schematic symbol style is broken.

## Change the Scale of the Drawing

5.  Change the drawing scale:

    a.  On the Desktop menu, click Drawing Setup.

    b.  Click the Scale tab.

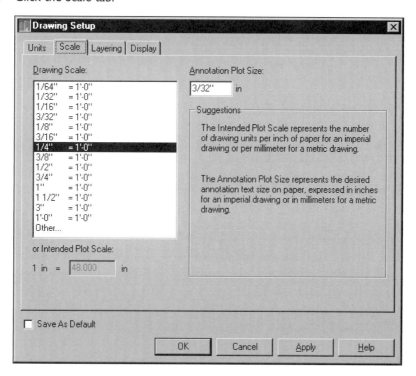

The Scale tab is where the annotation scale factor is found.

    c.  Change the annotation scale to 1/8", then click OK.

 **Note:** All the symbols in the drawing are scaled up a bit, except for the union.

    d.  Press CTRL+Z to undo the annotation change.

    e.  Type OBJRELUPDATE and then ALL at the command line to reestablish the relationships of the symbols and lines.

 **Note:** The purpose of this exercise is to show the behavior of the Use Annotation Scale check box on the Views tab of the Schematic Symbol Styles dialog box when scaling symbols. The most efficient way to change the symbol scale while preserving dynamic scaling is to adjust the X, Y, and Z Override Scale values.

6. Change the scale of the three-way valve:

    a. Select the three-way valve, right-click, and click Edit Schematic Symbol Style.

    b. Click the Views tab and enter **.75** for the X, Y, and Z scaling size.

    c. Click OK.

This is what your finished drawing should look like.

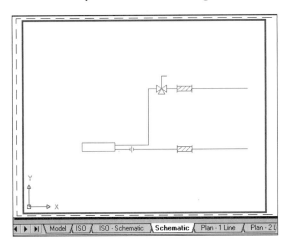

The drawing setup scale values of drawing scale and annotation plot size are used for many things in Architectural Desktop and Building Systems. Title mark and other annotation symbols from the DesignCenter use these values when they are placed in the drawing. The label for the schematic lines and symbols also use these values. Changing either of these values will affect all of these objects, not just the schematic symbols. Additionally, the drawing scale is directly linked to the dimension scale variable DIMSCALE. Be careful in adjusting these values.

OBJRELUPDATE is a command that is used to clean up the drawing. The command stands for Object Relationship Update. As you work, you may find that some commands cause the symbols to appear to lose their relationship with each other. While sometimes a REGENALL may clean up the drawing, at other times you may need to issue the OBJRELUPDATE command at the command line, or use the MEP Common menu to select Utilities ➤ Regenerate Model which does the same thing.

### EXERCISE 2: USING AUTOCAD COMMANDS TO MODIFY SCHEMATIC SYMBOLS

In this exercise you will be modifying your schematic drawing using basic AutoCAD commands. This exercise shows you the behavior when copying, moving, and erasing symbols that are anchored to lines, and lines anchored to symbols. Symbols anchored to lines will maintain their anchors and may be copied to any other schematic line with the same system. Symbols placed in the drawing first (those not anchored to the lines, but that may have lines anchored to them) will simply be copied, and will not copy the line with its anchored relationship.

 **Open:** *03 Moving Schematic Symbols.dwg*

### Copy and Move Schematic Symbols

1.  Move the union pipe fitting:

    a.  Select the left grip point on the union pipe fitting.

    b.  Drag your cursor to the left. You do not need Ortho turned on because the symbol is anchored to the line.

    c.  On the command line enter **3"**, and then press ENTER.

2.  Copy the union pipe fitting:

    a.  Select the union pipe fitting.

    b.  On the command line enter **copy selection** and then press ENTER.

    c.  Select a base point above the top connector line.

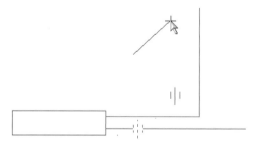

    d.  Click to place the union pipe fitting on the line above.

Notice that when you clicked to place the union pipe fitting that it snapped to the closest line.

3. Move the lower schematic line and flex connection pipe fitting:

   a. Select the lower line. Grips appear at the location points that were selected when the line was drawn. Hold the right mouse button down for a moment to deselect the line.

   b. Move the flex connection pipe fitting to the right 12". Moving the symbols also moves the lines anchored to it.

4. Erase a line that has an anchored symbol:

   a. Select the line that connects the coil to the upper flex pipe connection.

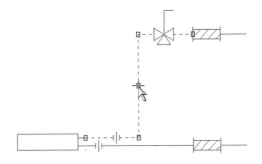

   b. Press the DELETE key.

Erasing a line that has symbols anchored to it also deletes all of the anchored symbols.

 **Note:** Enter **objrelupdate** and then enter **all** (for all objects), to refresh the appearance of the schematic symbol and schematic line.

This is what your finished drawing should look like.

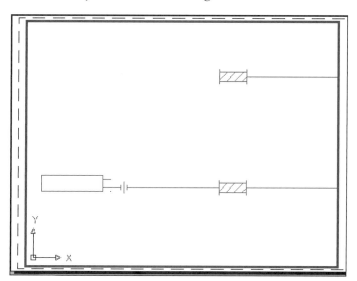

In Lesson 3, "Modifying your Mechanical Schematic Diagram," you completed two exercises: Modifying Schematic Symbol Properties and Scale and Using AutoCAD Commands to Modify Schematic Symbols. You have now successfully modified a mechanical schematic drawing that represents the piping schematic for a reheat coil for a VAV control box. In the next lesson you will continue to work on the schematic diagram. You will create schematic systems, and custom schematic line and symbol styles.

## LESSON 4: CREATING SCHEMATIC LINE STYLES, SYSTEM DEFINITIONS, AND SYMBOL STYLES

This lesson contains three exercises. Each exercise has a drawing file associated with it. When you complete the exercises in this lesson you will have created Schematic Line Styles, System Definitions, and Symbol Styles. Styles are stored in the drawing similar to a block, and may be copied from one drawing to another. The interface for storing and creating styles is Architectural Desktop's Style Manager. The Style Manager allows you to access the definitions for styles in the drawing. Again there is some terminology here that is a bit confusing. For the Schematic Line, there is a schematic line style that controls what is displayed when lines cross. There is also a schematic system definition that controls the layer the line is placed on. Both the styles and the definitions are accessed through the Style Manager. In the exercises that follow, you use Style Manager to create new styles and to access the style definitions for editing. In the final exercise you use Style Manager to copy the schematic symbol style that you created into a drawing stored on your hard drive so that you can access it from the Add Schematic Symbol dialog box.

A schematic symbol is a style-based object. Nested within the symbol is a standard AutoCAD block. Due to the fact that schematic symbols are style-based objects, schematic symbols have the ability to contain information such as a general description, how the symbol should break into a line, the amount and locations of connections that should be placed on a symbol, and the ability to assign different blocks to different views. You can also assign symbol properties for labeling. Lines are also style-based objects, and have similar built-in functionality. System Definitions apply to both symbols and lines, and are yet another style stored in the drawing. The system definition controls the layering of the line or symbol. The system also contains the system abbreviation used in labeling.

To accomplish this task you will be using Architectural Desktop's Style Manager, the Schematic Line Styles dialog box, the Schematic Symbols Styles dialog box, and a pre-drawn AutoCAD block. At the end of each exercise there is a screen shot of what your finished drawing should look like.

This lesson contains the following exercises:

- Schematic Line Systems and Styles
- Creating Schematic Symbols
- Reusing Styles

### EXERCISE 1: SCHEMATIC LINE SYSTEMS AND STYLES

In this exercise you will be creating schematic System Definitions to represent heating water supply and heating water return piping. During this process you will be using Style Manager to create the new styles, along with the schematic system definition and the Schematic Line Styles dialog boxes.

In this exercise you will create two schematic system definitions to control the layer and linetype of the lines. You will also create a schematic line style to control how lines break when they cross.

 **Open:** *03 Line Styles and Systems.dwg*

#### Create Heating Water Supply and Return Schematic System Definitions

1. Access Schematic System definitions:
    a. From the MEP Common Menu click Schematics ➤ Schematic System Definitions.
    b. Select New Style icon (not the new drawing icon), right-click on the new style, and click Edit.

2. Specify description and annotation information for the new System Definition:

   a. Click the General tab, enter **HTWS** for name, and enter **Heating Water Supply** for description.

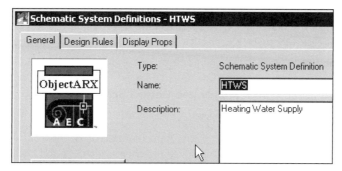

   b. Click the Design Rules tab.

   c. Enter HTWS for the abbreviation to be used with the label.

   d. Pick mhotw for the layer key style. This will cause lines and symbols of this system to be placed on the layer M-Hotw for the heating water system.

   e. Pick the ellipse next to the minor button.

   f. Select Supp to add an override to the layer name, this way the final layer name for either symbols or lines that are assigned this system will be M-Hotw-Supp.

   g. Enter **HTW** for the System Group. The System Group HTW is already assigned to the Equipment Schematic System Definition. Assigning the same System Group to the different system allows the connections between these systems to be valid.

3. Specify the display properties for the new HTWS system definition:

   a. Click the Display Props tab, under Property Source select Schematic System Definition, and then click Attach Override.

   b. Click Edit Display Props to access the Entity Properties dialog box.

   c. Change the Schematic Line color to red and the linetype to continuous, and then click OK twice to get back to Style Manager.

   Change the values in this dialog box by clicking on them.

4. Create another system definition by copying the HTWS style:

   a. Select HTWS, right-click and click copy and then right-click and click paste.

   b. Select HTWS (2), right-click, and click Edit.

5. Specify description, annotation, and Display information for the HTWS (2) style:

   a. Click the General tab, enter **HTWR** for name and enter **Heating Water Return** for description.

   b.   Click the Design Rules tab, enter **HTWR** for the Abbreviation, change the Minor override to Retn, and verify the System Group is HTW.

   c.   Click the Display Props tab, under Property Source select Schematic System Definition, and then click Edit Display Props.

   d.   Change the Schematic Line color to blue and the linetype to Hidden2 and then click OK twice to get back to Style Manager.

   e.   Click OK in Style Manager to return to the drawing.

### Apply the Heating Water Supply and Return Schematic System Definitions

6.   Change a line's System to HTWR:

   a.   Select the line coming from the upper connector of the coil, along with the lines to the right of the valve, and then right-click and click Schematic Line Modify.

   b.   Change the System to HTWR and then click OK.

7.   Change the remaining lines Systems to HTWS.

The System Style controls the looks of all lines in that system. The System Group determines if connections are valid between objects on different systems. In this case, the 3 way valve using the equipment system did not clean up with the lines assigned the standard system. Once the lines were assigned a system with the same system group, the lines are hidden by the 3 way valve.

Separate from the Schematic System Definition, is the Schematic Line Style. The Schematic Line style controls how the lines break when they cross. In the next section of this exercise you will create a Schematic Line style.

### Create a Copper Schematic Line Style

8.   Access schematic line styles:

   a.   From the MEP Common menu, click Schematics ➤ Schematic Line Styles.

   b.   Select New Style, right-click, and click Edit.

9.   Specify description and annotation information for the new schematic line style:

   a.   Click the General tab, enter **Copper** for name, and enter **Copper Piping** for description.

   b.   Click the Designations tab. Multiple designations are allowed for schematic lines. We will use this as a storage place to hold some size designations that can be used later for labeling.

   c.   Pick the new designation icon.

   d.   Enter **1/2"**.

e.    Repeat adding **3/4"**, **1"**, **1 1/4"**, and **1 1/2"**.

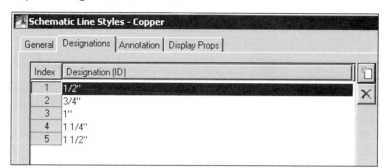

f.    Click the Annotation tab and under Crossings select Break Existing Line.

g.    Verify the Gap/Graphics Width is **1/16"**.

h.    Verify that the Connections as well as the Start and End Settings are set to None. (Note the priority can be used to control which line cuts through which when more than one line style is used in a drawing.)

i.    Click OK twice to get back to the drawing.

### Apply the Schematic Line Style

10.  Assign the line style to the lines:

a.    Select all the lines, right-click, and click Line Modify.

b.    Change the style to copper.

c.    The gap shows up where the two lines cross with the supply line breaking the return line.

d.    Select Tools ➤ Display Order ➤ Send to back and pick the supply line at the crossing to make the return line break the supply line.

This is what your finished drawing should look like.

Both the Schematic Line style and the Schematic system Definition have a display properties tab. You can control layer/color/linetype with either of these definitions. We suggest you pick one, and use that exclusively. This will eliminate one variable you have to keep track of when you are trying to work with a drawing created by someone else in the office.

As we have illustrated with this exercise, you can change the properties of the Schematic line, the starting block, or the ending block. These are the subcomponents of the schematic line object. The default setting for both the schematic line and the schematic system for all the subcomponents is layer 0, color = ByBlock, Linetype = ByBlock. With these settings, the line will inherit the settings of the layer they are placed on, which is reasonable.

The Display properties are a basic part of most Building Systems objects. When I am setting up styles or system definitions, sometimes I will need a subcomponent to be a different color than the main object (in the case of the schematic line, I might want the start and end blocks to be a different color than the line). In these situations, I will usually create a new layer for the subcomponent to be on, and then modify the style or system definition's display properties, and pick on the layer 0 and assign the subcomponent to the new layer, keeping the color and linetype ByBlock. This gives me a couple of results: If I give my drawings to someone with Object enabler, they will not have access to the Style Manager, but will be able to use standard layer control to change the color/linetype of the subcomponents that are buried in the AEC object itself. Secondly, if I need to explode the object for whatever reason, the subcomponents will explode down to their assigned layers and not layer 0.

### EXERCISE 2: CREATING SCHEMATIC SYMBOLS

In this exercise you will be creating a schematic symbol to represent a water source heat pump. To create a schematic symbol you need to draw lines to create a block as the basis for your schematic symbol. You then assign a style, connectors, and scale to the block in order to create a custom schematic symbol.

In this exercise, a block named WSHPSchematic has already been drawn for you. You will be using this block as your basis for creating a schematic symbol. The block is a standard AutoCAD block with the entities on layer 0 with color and layer attributes pre-assigned by layer or by block. The block is 1" high and 2" long, to stay standard with other HVAC equipment. You will be adjusting the scale of the block when you create the schematic style.

 **Open:** *03 Creating Schematic Symbol Styles.dwg*

## Create a New Schematic Symbols Style

1. Access schematic symbol styles:

   a. From the MEP Common Menu click Schematics ➤ Schematic Symbol Styles.

   b. Select Schematic Symbol Styles, right-click, and click New.

   c. Select the New Style, right-click, and click Edit.

2. Specify description and line clean up rules for the new symbol style:

   a. Click the General tab, enter **WSHP** for name, and enter **Water Source Heat Pump** for description.

   b. Click the Design Rules tab and click Trace Geometry. The line cleanup rules tell the symbol how to hide the schematic lines as they are connected to the symbol, or the symbol is placed on a schematic line. Trace geometry is the default for most parts, and will hide the schematic line as it touches the geometry of the block.

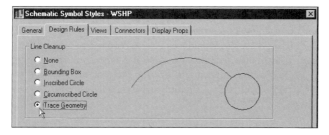

## Assign the View Blocks and Scale to the Symbol Style

3. Specify the view directions for the symbol:

   Schematic symbols are comprised of multi-view blocks and therefore can have different appearances in different view directions.

   a. Click the Views tab and click Add to add a view.

   b. Verify that the mode is set to Ortho.

   c. Change the View Name to **schematic,** and in the View Block pull-down select WSHPSchematic.

   You can assign different AutoCAD blocks as view blocks to this symbol. Here we are creating a symbol that uses the same view block in all views. Additionally, this symbol is created for orthogonal mode.

   If you are creating a schematic symbol that will be used for isometric diagrams, you will need to create a group of blocks that will be read into the proper orientation with respect to the line. If you want to see how they are put together, use the Style Manager to open the drawing *C:\Program Files\Autodesk Building Systems 3\Aecb Content 3\Schematic\Plumbing Symbols - Isometric.dwg* and look at the fixture stacks created there.

d.  Select Use Annotation Scale and Override Scale and then change the *X, Y,* and *Z* override values to **4**.

  **Note:** Most symbols use 2.275 as the scale override, therefore, this symbol will be larger then the other symbols as it is inserted.

Originally designed for annotation such as the labels and title marks, the annotation scale factor applies to symbols as well. The annotation scale is accessed from the Desktop ▸ Drawing Setup, Scale tab. The annotation scale uses the annotation factor and the drawing scale together. The original block is 1" x 2". When this symbol is inserted, a drawing scale of ¼" = 1'-0" will multiply this by 48 so that the size is 4'x8'. The annotation scale 1/8" then multiplies the symbol by 1/8 so that the size is 6"x12". If this were text, you can see where this would make sense. You would create all of your annotation blocks with 1" attributes. The drawing scale and annotation scale would then take care of sizing the text to the appropriate size for the plot sheet. While this setting is not dynamic for text blocks such as the title marks, it is dynamic for the schematic symbols and the labels.

**Add Connectors to the Symbol Style**

4.  Create the return and supply air connectors for the symbol:

All Building Systems objects have connectors. Each connector that you define has a connector type. Each connector type has a domain, description, and connection points. The default for connection points for symbols is at the start and end points of the symbol. If you wanted the connection point to be at the center of the symbol, then you would click Add Center Point from View Blocks in Top View.

a.  Click the Connectors tab and click Add to add another connector.

b.  Change the Description to **AirSupply**, and under Connection Point enter **.5** for *X*, **1** for *Y*, and **0** for *Z*.

c.  Click Add, change the Description to **AirReturn**, and under Connection Point enter **.5** for *X*, **0** for *Y*, and **0** for *Z*.

5.  Create the piping connectors for the symbol:

a.  Click Add, change the Description to **WaterSupply**, and under Connection Point enter **2** for *X*, **1/8** for *Y*, and **0** for *Z*.

b.  Click Add, change the Description to **WaterReturn**, and under Connection Point enter **2** for *X*, **1/4** for *Y*, and **0** for *Z*.

c. Click Add, change the Description to **Condensate**, and under Connection Point enter **2** for *X*, **1/2** for *Y*, and **0** for *Z*.

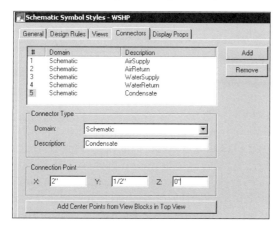

d. Click OK to get back to Style Manager.

6. Review your new schematic symbol:

a. Click the WSHP style.

You can see how your symbol will be displayed in the drawing on the Viewer tab (you may have to zoom out to see the entire symbol).

b. Click the List tab to verify the information you defined for the symbol.

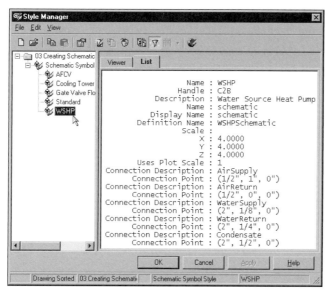

c. Click OK to get back in the drawing.

7. Save the drawing symbol:

   Once you have created a drawing symbol, you can re-use it in other drawings.

   a. From the File menu, click Save As.

   b. Browse to *C:\Program Files\Autodesk Building Systems 3\Aecb Content 3\Schematic.*

   c. Enter **Exercise Examples.dwg** for the File name and then click Save.

   d. Close this drawing if you are proceeding with the next exercise.

## EXERCISE 3: REUSING STYLES

In this exercise, you will be reusing the schematic symbol style that you created in exercise 2 that represented a water source heat pump. In order to accomplish this task, you will be using Style Manager to copy and paste schematic styles from one drawing to another. This saves time in creating new styles. You can also use a style and modify only certain properties of that style in order to tailor it to your needs. In this drawing, you will use the Style Manager to open the drawing created in the last exercise.

 **Open:** *03 Sample Schematics.dwg*

### Use the Style Manager to Access Styles in a Different Drawing

1. Open the Exercise Examples drawing in Style Manager:

   a. From the MEP Common Menu, click Schematics ➤ Schematic Symbol Styles.

   b. From the File menu in Style Manager, click Open Drawing.

   c. Browse *C:\Program Files\Autodesk Building Systems 3\Aecb Content 3\Schematic* and select the Exercises Examples drawing created in the last exercise.

   d. Click Open.

 **Note:** When you open a drawing from the Style Manager it does not open in the AutoCAD session. It is open to the Style Manager only to allow you to copy styles from one drawing to another.

2. Access the schematic symbol styles:

   a. In the left panel of Style Manager, expand the *03 Sample Schematics.dwg* folder and the *Exercises Examples.dwg* folder.

   b. Expand the Schematic Symbol Style option in both drawings.

## Copy Styles Between Drawings with Style Manager

3.  Copy schematic symbol styles:

    a.  Select the individual Schematic Symbol Style Large tank in the *03 Sample Schematics.dwg*, right-click, and click Copy.

    b.  Select the Schematic Symbol Style in the *Exercises Examples.dwg*, right-click, and click Paste.

        View the schematic symbol styles that you now have available to you in the *Exercises Examples.dwg*.

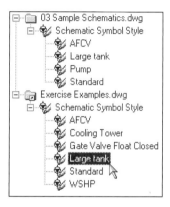

4.  Drag-and-drop schematic symbol styles:

    a.  Highlight the Schematic Symbol Style in the *03 Sample Schematics.dwg*. You can view the schematic symbol styles available in the drawing in Style Manager.

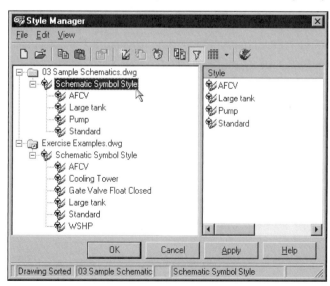

b. In the left panel, select the individual schematic symbol style Pump and drag it down into the Schematic Symbol Style in the *Exercises Examples.dwg*.

c. Click OK to close the Style Manager.

d. Click Yes to save the changes to the *Exercises Examples.dwg*.

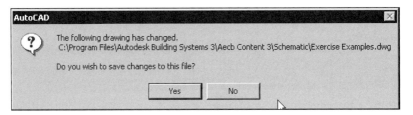

**Note:** Even though the Exercise Examples.dwg was not open to AutoCAD, but just to the Style Manager, you have changed this drawing by adding the new styles to it. You need to save these changes to the *Exercise Examples.dwg*.

These schematic symbol styles now exist in the *Exercises Examples.dwg*. They are available the next time you use the Add Schematic Symbols command. The next part of this exercise illustrates how to group the schematic symbol styles together within the *Exercise Examples.dwg*.

### Create a Symbol Category and Organize Symbols

5. Open the *Exercise Examples.dwg* and access the schematic symbol styles:

a. Close the *03 Sample Schematics.dwg*. The Building Systems session should have no drawings open at this point.

b. Open the *Exercises Examples.dwg* in the *C:\Program Files\Autodesk Building Systems 3\Aecb Content 3\Schematic* folder.

c. From the MEP Common Menu, click Schematics ➤ Add Schematic Symbol.

d. In the Schematic Symbol drop-down, select Current Drawing.

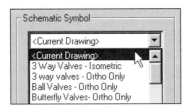

6. Categorize the schematic symbols:

a. Right-click on the All bar and click New Category.

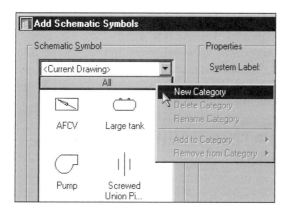

b.   Enter **Heating Water** for the New Category name.

7.   Add symbols to the new schematic symbol category:

a.   Right-click on the Pump symbol and click Add to Category ➤ Heating Water.

b.   Right-click on the Water Source Heat Pump symbol and click Add to Category ➤ Heating Water.

c.   Click the Heating Water bar and view the symbols that you have added.

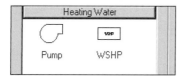

8.   Use the schematic symbols in another drawing:

a.   Save and close the *Exercises Examples.dwg*.

b.   From the File menu, click New.

c.   From the MEP Common Menu, click Schematics ➤ Add Schematic Symbol.

d.   Use the drop-down to change the symbol list to Exercise Examples.

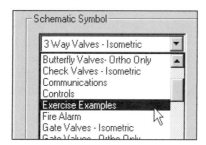

Notice that the symbols and categories are available for you to use.

In Lesson 4, "Creating Schematic Line and Symbol Styles," you completed three exercises: Creating Schematic Line Styles, Creating Schematic Symbols, and Reusing Styles. You have now successfully created schematic line and symbol styles and have copied the schematic symbol styles that you created into a drawing. In the next lesson you will be annotating the heating water system that you drew in lesson 2 and modified in lesson 3.

## LESSON 5: ANNOTATING YOUR MECHANICAL SCHEMATIC DIAGRAM

Building Systems works in conjunction with Architectural Desktop to supply you with the annotating features that you need. The annotation provided can be generalized into two types of annotation. Building Systems offers style-based annotation in the form of Labels. Architectural Desktop provides a group of annotating functions that are launched from the DesignCenter.

Labels are style-based annotation objects that will read information stored on the AEC objects. When a label is placed in the drawing, it becomes anchored to a schematic line or symbol. The label reads a value from the object, which it then displays as text in the drawing. What aspect of the object is read into the label is determined by the settings in the label style. The label's style also controls dimensions, text options, and display properties of the text. There are several label styles in the Aecb Building model template:

- The System Label style displays at the system abbreviation stored in the schematic system assigned to the line or symbol.
- The Name style of label reads the style name of the symbol or the style name of the schematic line.
- The Standard style of label reads the designation ID from the line or symbol.

Labels, like schematic symbols, are style-based and are able to read the information stored in the schematic line and symbol styles. Architectural Desktop supplies you with documentation symbols such as title marks, leaders, and detail marks.

How the text is formatted is determined by variables held in the AutoCAD text style that is used by the label. If the AutoCAD text style has a height set, the label will use that text height for the label text absolutely and not use the annotation scaling factor. If the AutoCAD text style has a height set to 0", then the label will use the annotation plot size in the drawing setup to determine the text height.

Labels can have three different modes. The modes determine how the labels are repeated along a line. In these exercises, you will use Manual spacing where you are prompted for the location of the label. You may also use the Space Evenly option, which will place a given number of labels equally along the schematic line, or Repeat, which will place a given number of labels at a given distance between each label.

 **Note:** The drawings in this exercise have been redrawn using different scaling factors for this lesson. Please use these drawings and not drawings from previous exercises.

The second type of annotation is borrowed from Architectural Desktop. These commands are accessed from the Documentation menu. Title marks, section marks, and revision clouds are some of the annotations that can be added to a drawing from the DesignCenter. These annotations all launch a series of commands when they are added to the drawing. Be sure to pay attention to the command line when adding these annotations.

While these annotations will utilize the annotation and drawing scale of the drawing setup when they are originally placed in the drawing, changing either of these settings after they have been placed will not affect their size.

This lesson contains two exercises. Each exercise has a drawing file associated with it. When you complete the exercises in this lesson, you will have annotated the heating coil schematic.

To accomplish this task you will be using the Building Systems labeling command and Architectural Desktop's documentation content, which are accessed through AutoCAD's DesignCenter. At the end of each exercise there is a screen shot of what your finished drawing should look like.

This lesson contains the following exercises:

- Labeling your Schematic Drawing
- Adding Documentation Symbols to your Schematic Drawing

### EXERCISE 1: LABELING YOUR SCHEMATIC DRAWING

In this exercise you will be adding labels to a schematic heating water system. During this process you will be using the labeling commands included with Building Systems.

 **Open:** *03 Adding Labels.dwg*

### Add Labels with Different Styles

1. Add a label to the heating water supply line:

    a. From the MEP Common Menu, click Labels ➤ Add Label.

    b. Pick the lower line coming from the bottom coil connector.

    c. At the command line, enter **system label** for the style name and then press ENTER three times to accept the defaults.

d. When the command line says Node Position, pick the lower line coming from the bottom coil connector.

 **Note:** The label is anchored to the line and reads the HTWS from the abbreviation variable held in the schematic system.

2. Add a label to the coil:

   a. From the MEP Common Menu, click Labels ➤ Add Labels.

   b. Pick the coil.

   c. At the command line, enter **standard** for the style name and then press ENTER three times to accept the defaults.

   d. When the command line says Node Position, pick the coil.

When you label a symbol, the label will be placed at one of the connectors.

## Modify the Labels

3. Reposition the coil label:

   a. Pick the coil label, right-click, and click Offset Node.

   b. Pick a point above the coil to reposition the label.

4. Edit the style to force the label horizontal:

   a. Pick the coil label, right-click, and select Edit Label Style.

   b. On the Label Style Offset tab, check the box next to Force to Horizontal Justification.

This is what your finished drawing should look like.

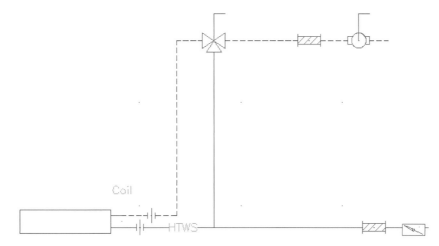

### EXERCISE 2: ADDING DOCUMENTATION SYMBOLS TO YOUR SCHEMATIC DRAWING

In this exercise you will be adding a title mark and a leader to a schematic heating water system. During this process you will be using Architectural Desktop's documentation symbols, which are accessed through AutoCAD's DesignCenter. The DesignCenter starts commands as you drag content such as titles and leaders into the drawing. The content that is placed in the drawing is standard AutoCAD text, lines, and leaders. The drawing setup scale determines the scale of the documentation symbols. These commands read and use the current AutoCAD text style.

 **Open:** *03 Adding Titles and Notes.dwg*

**Add a Title Mark**

1. Access Architectural Desktop's Title Marks:

   a. From the Documentation menu, click Documentation Content ➤ Title Marks.

   Notice that the left panel of DesignCenter opens to the Title Marks folder and all of the available title mark styles are displayed in the right panel.

   b. In the right panel of DesignCenter select Title Mark A1 and drag it into the drawing.

   c. Pick an insertion point under the coil.

2. Place the title mark in your drawing:

   a. Click OK to accept 1 for Number in the Edit Attributes dialog.

b.   Enter **Not to Scale** for scale and enter **Coil Supply and 3 way valve 1/2" - 2"** for Title, and then click OK.

c.   Drag your cursor to the right and click at the end of the title to underline the title mark.

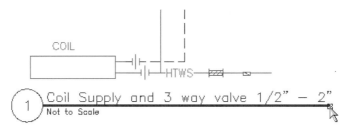

**Add a Leader**

3.   Access Architectural Desktop's Leader lines:

a.   From the Documentation menu, click Documentation Content ➤ Leaders.

b.   In the right panel of DesignCenter, select Straight [Text] and drag it into the drawing.

c.   Hold the SHIFT key and right-click to get the nearest OSNAP to help pick a starting point of the leader (where the arrow will point) on the vertical heating water supply line.

4.   Place the leader line in your drawing:

a.   Pick a second point to determine the vertex of the leader line.

b.   Pick a third point to determine the end of the leader line where text will begin.

c.   Press ENTER to start placing text.

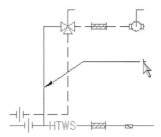

d.   On the command line, enter **FULL SIZE** for the first line of text.

e.   Press ENTER, enter **BY-PASS** for the next line of text, and press ENTER twice to end the command.

This is what your finished drawing should look like.

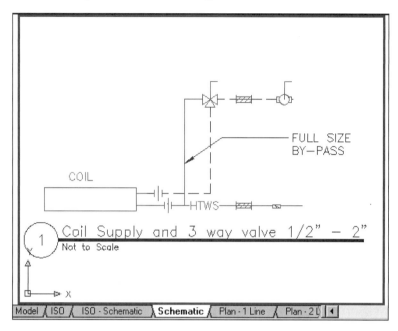

The Title and the leader are sized according to the variables held in the drawing setup. All of the annotation that is used from the documentation pulldown will use the current text style as the style of the text placed in the drawing. Set your office standard text to current before you start annotating the drawing using these commands.

Congratulations! You just made a schematic diagram!

## KEY CONCEPTS: SCHEMATIC DIAGRAMS

Use the schematic tools to create preliminary layouts or final diagrams.

Schematic symbols and lines may be drawn in isometric or orthogonal mode, but it's better not to mix the two in the same diagram.

Schematic Symbols and Schematic Lines are style-based objects.

The object placed in the drawing first will be the controlling object, and the object added to it will be the anchored object.

The drawing and annotation scale of the drawing setup affect the size of schematic symbols.

The drawing and annotation scale of the drawing setup affect the size of labels unless the label style uses an AutoCAD text style with a fixed height.

Schematic lines styles hold the control over how the lines break when they cross.

Schematic systems hold the layer keys used by a line or symbol assigned to them.

Schematic symbols styles can be copied with Style Manager into other drawings.

Schematic symbols may be stored for access in the folder pointed to by the Building Systems Catalog tab of the Options dialog box.

# CHAPTER 4

# Creating a Mechanical Plan

After you read this chapter, you should understand the following:

- how to specify the drawing options and preferences for drawing ducts and equipment
- how to place equipment for the mechanical system
- how to route ductwork through the building to connect the equipment
- how to modify the design
- how to work efficiently with HVAC system definitions and layer keys

## CREATING A MECHANICAL PLAN

When you draft a mechanical plan with Building Systems, you will be working with objects. Like the schematic symbols and schematic lines from the last chapter, you will use Building Systems objects to represent the items in the mechanical plan. These objects are added to and behave in the drawing very differently than standard AutoCAD entities such as lines arcs and circles. The objects you use to create the mechanical plan are Multi-View Parts (MvParts), ducts, and fittings. First, you add the MvParts to represent the fans, pumps, air terminals, and other equipment. After the equipment is in place, you will connect the parts with ducts. As you draw ducts, the fittings such as elbows, tees, and taps are automatically placed in the drawing for you.

## THE BUILDING MODEL

When you add ducts, you are creating the building model, not just drawing lines to represent the equipment and ducts. The background drawings associated with the exercises have been created in Architectural Desktop. As objects, the walls, doors, stairs and structural members can be viewed as a 3D building model. Working in 3D will take some adjustment for those of you who have never done so before. However, Building Systems provides many tools to simplify working in 3D. Much of the 3D navigation is done in plan view using preset elevations that place objects at their proper Z-coordinates. The Building Systems compass tool aides in creating ducts through the building model. With auto-routing, you can pick two non-aligned points in the drawing and have the software give you a series of duct routes to choose from.

Working in 3D is not the only difference in working with Building Systems. The process of getting from a schematic system diagram to finished construction documents is different from working in plain AutoCAD. Early on in the design, you may not have all of the 3D information available to create a full building model. The duct object has an "undefined" setting that allows you to place a single line duct in the drawing without specifying the size or system of the duct until later when these values have been determined. Once the system has been designed, you can go back and convert the undefined ducts into duct objects that have a size, system, and all of the display representations: one-line, two-line, and model. When working with the undefined type of duct, don't worry about setting an elevation, just draw the duct in the plan. Later, when you add the size and system information to the undefined ducts, you can set the elevations as well. Once the horizontal trunks are in the drawing, you can add the elevation transitions to the system.

## MULTI-VIEW PARTS (MVPARTS)

Briefly described in chapter 1, Multi-View Parts (MvParts) represent the equipment in the drawing. These pieces of equipment are stored in catalogs and added to the drawing as needed. A simplistic view of an MvPart is a collection of blocks with connectors at locations where ducts, pipes, or conduit will be attached. Each view direction of a display representation of the MvPart can be assigned a different block.

This allows the MvPart to be shown differently in a schematic or one-line plan than a two-line plan. It also allows the MvPart to be displayed as a 3D block when seen in Isometric mode and a 2D block when seen in the plan view.

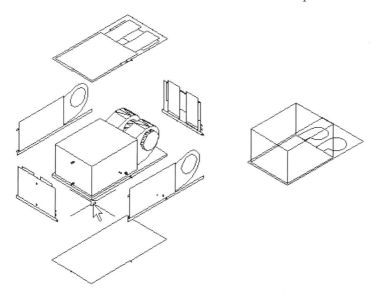

Connectors are built into MvParts. At each location on the MvPart where a duct or pipe will attach, there will be a connector. The connector holds the size and shape of the duct that will be attached to that location. The connector can also hold an HVAC system. As you add ducts to a connector on an MvPart, the duct will automatically size to the connector size. The duct will also assume the HVAC system from the connector if one has been assigned to it. By default, when you place an MvPart in the drawing, each of the connectors of the part is assigned the system "NONE", allowing you to add ducts of any system to that connector. You may manually change the system assigned to a connector to control what system of duct will be attached to it.

MvParts are stored in a catalog that is installed with the software. When you add an MvPart, the part's style is based on its definition in the catalog. For each part, several sizes may be used in any drawing. Each different size of the same part creates its own style. The second lesson in this chapter contains information about adding MvParts to the drawing.

## DUCTS

With the equipment MvParts in place, the next task will be to add ducts. When you add ducts you will be routing them through the building model. This chapter focuses on adding ducts, illustrating the different ways the software allows you to route the ducts around objects. You can manually route ducts through different elevation changes, or pick a start and endpoint and let the software route the duct from point to point. The exercises in lesson 3 show you how to add ducts to the building model.

## LAYER KEYS

Each Building Systems object has a built-in, or native, layer key. Within every Building Systems drawing is a layer key style. The layer key style is like a spreadsheet that links each object with a set of layer settings. When an object is added to a drawing, a string of events happens. First, the object looks in the layer key style for the layer associated with its layer key. Then it looks to the existing layer list in the drawing. If the layer exists, no changes are made. If the layer does not exist, the layer is created with the settings from the layer key style.

For example: If you add a duct, it looks into the layer key style for its layer key *Duct*, and finds the layer M-STND-Duct. Next, the list of currently defined layers is matched against the layer name M-STND-Duct. If the layer M-STND-Duct exists in the drawing, but its color is yellow and linetype is dashed, the duct will use this layer as it is and assume its yellow and dashed properties. If the layer M-STND-Duct does not exist in the drawing, a new layer M-STND-Duct would be created. Its color will be 112 and linetype continuous as defined by the layer key style for the layer key *Duct*.

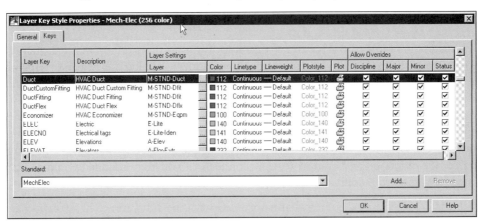

Every Building Systems object has a layer key. The source of the layer key may be hard coded by the software, as in the case of ducts and fittings, or stored in the parts catalog, as are the layer keys for the MvParts. Lesson 5 discusses some of the controls for managing the layering of building systems objects. In this lesson you will find exercises on global overrides that affect all new objects you place in the drawing. This lesson also contains exercises on using the system definitions to control the layers of the objects.

## SYSTEM DEFINITIONS

In the last chapter, when you worked with schematic lines and symbols, you were introduced to the schematic system definition. MvParts, ducts, and fittings are also assigned a system definition; in this case, an HVAC system definition. Both types of system definitions may include a layer key, system abbreviation, and system group, along with display overrides that allow each system's subobjects to have a unique appearance in the drawing. However, system definitions are applied to the schematic objects differently than MvParts, ducts, and fittings. Lesson 5 leads you through four exercises in which you create and apply HVAC system definitions to the mechanical model.

## LESSON OBJECTIVES

This chapter contains five lessons that illustrate how to create and work with the MvParts and ducts. Each lesson is broken down into exercises. Most exercises have a corresponding drawing file that is located on the CD-ROM included with this book.

Lessons in this chapter include:

- **Lesson 1: Specifying Your Drawing Options**
  In this lesson you review the background settings and environment variables in the Options pages and Duct Layout Preferences. These settings will affect how the software behaves.

- **Lesson 2: Placing the Equipment**
  In this lesson you add MvParts to your drawing. There are exercises to create the air handler and add return grills and air diffusers.

- **Lesson 3: Working with Ducts and Fittings**
  In this lesson you route ductwork through the building to connect the equipment and air terminals.

- **Lesson 4: Modifying Your Mechanical Plan**
  In this lesson you change the sizes of the ducts and equipment.

- **Lesson 5: Working with System Definitions and Layer Keys**
  In this lesson you create an HVAC system definition and assign it a layer key. You also modify the system assigned to a duct and a piece of equipment. You will use a layer key override to assign a layer modifier as you draw new objects.

## LESSON 1: SPECIFYING YOUR DRAWING OPTIONS

Like AutoCAD, Building Systems gathers many environmental variables together and provides access to them through the Options dialog box. For example, you can control how different types of ducts connect and how overlapping ducts are displayed through the Building Systems tabs of the Options dialog box. In addition to these settings, the duct layout preferences control how a duct will be added to the drawing and what fittings will be used by default at corners and intersections of ducts. The settings and variables in these exercises are stored either in the drawing, or in the registry of your computer. Most of the variables used for the exercises in this book use the default settings installed with the software.

This lesson contains the following exercises:

- Specifying the Building Systems Options
- Specifying the Duct Layout Preferences

The first exercise leads you through the five tabs of the Options dialog box that pertain specifically to building system's objects. The second exercise walks you through the Duct Layout Preferences dialog box and the alternatives you can set here to determine how ducts are added to the drawing. Each of these exercises has a drawing associated with it. When you complete the exercises in this lesson, you will have reviewed the background variables for the software.

### EXERCISE 1: SPECIFYING THE BUILDING SYSTEMS OPTIONS

The five tabs in the Options dialog box specifically relate to the Building Systems software. The Layout Rules tab is a catch-all for different conditions that occur when you add ducts to the drawing. The Crossed Objects tab has controls to tell the display system how to show ducts that overlap each other at different elevations. The Elevations tab is where you can add preset elevations to use when you add MvParts, ducts, or pipes. The paths to the content are stored on the Catalog tab. Lastly, the Building Systems Tooltips tab toggles on or off various cursor-activated tips that help you keep track of the objects in the drawing. In the following exercise, you will review the settings in each of these tabs. You will not be modifying each of the settings, but a brief explanation is given for what each of the settings do.

 **Open:** *04 Setting Building Systems Options.dwg*

### Access the Building Systems Tabs in the Options Dialog Box

1. Open the Options dialog box:
   a. From the Tools menu, click Options.
   b. Use the arrows in the upper-right corner of the dialog box to scroll to the right until you see the Building Systems tabs.

2.  Review the layout rules:

    a.  Click the Building Systems Layout Rules tab.

    b.  Verify the settings are as shown.

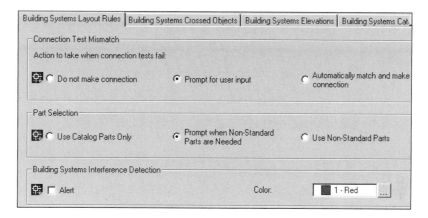

The Connection Test Mismatch determines what happens when you attempt to connect two objects together; for example, a duct on the return system to a duct on the supply system. The Part Selection area determines what happens if you attempt to add a non-standard part, such as a duct with a fractional dimension or a dimension larger than is defined in the catalog. Parametric parts, such as ducts and fittings, can be created in the drawing with sizes that are not specified in the catalog. The third area on this tab is a toggle that turns the interference detection on or off. You will look at this setting in a later exercise.

The next tab, the Crossed Objects Settings, controls how ducts are displayed when they overlap or cross each other. There are two different settings on this tab: "Using Haloed Lines Representation" and the "Using Hide Command" settings. These settings are mutually exclusive because they affect different display representations of the ducts or pipes. In the Aecb Building Model (Imperial) 3 template, the haloed line representation is used in the "Plan-2 Line" layout, and the 2-Line representation is used in the "Plan 2 Line-RCP" layout. In the next part of the exercise you will modify the settings and compare these two different layouts.

**Modify the Crossed Objects Settings**

3. Change the Crossed Object settings for the 2-line layout:

   a. Click the Building Systems Crossed Objects tab.

   b. Check the Apply Gap to Inside.

   c. Change the B – Gap to 2".

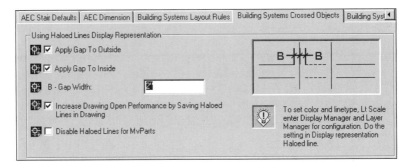

   d. Click OK to return to the drawing.

With the settings you specified in the Crossed Objects tab of the Options dialog box, the duct below is hidden by the upper duct.

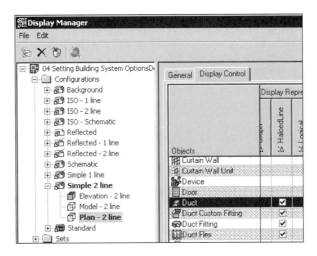

The previous image is a snapshot of the Display Manager showing the settings that are used for this viewport in this layout. This viewport has been assigned the display configuration "Simple 2 line." The "Simple 2 line" configuration uses the display set "Plan–2 line" when seen from the top view. This representation set uses the Haloed Line representation of the duct.

4.  Click the Plan 2 line–RCP layout tab.

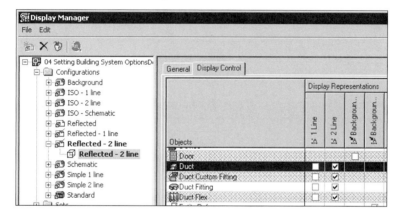

The Plan 2 line–RCP layout tab has a viewport assigned the Reflected–2 line display configuration that uses the 2 Line representation of the duct. This representation is different than the haloed line representation, and therefore does use the values you specified for the hidden line representation.

**Adjust the Crossed Objects Hidden Line Settings**

5.  Open the Options dialog box:

    a.  From the Tools menu, click Options.

    b.  The Building Systems Crossed Objects tab should be current.

6.  Adjust the settings:

    a.  Check the Show Crossed Objects as box.

    b.  Use the drop-down list to change the crossed objects setting to Dashed.

    c.  Check the Hide Lower Crossed Object box.

    d.  Leave the default 10 for the Break Width.

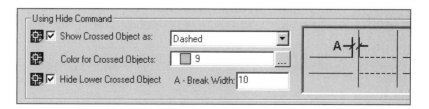

7. View the changes:

    a. Click OK to return to the drawing.

    b. At the command line, enter **Hide** to hide the lines.

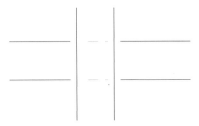

    The display controls of the Hide command are not as adjustable as the crossed objects controls for the haloed line representation.

## Add Elevations to the Drawing

8. From the Tools menu, click Options.

9. Add Elevations:

    a. Click the Building Systems Elevations tab.

    b. Click the Add Elevation icon in the lower-left corner of the dialog box.

    c. For the New Elevation Name, enter **Mechanical floor**, and then enter **10'-0"** for Elevation.

    d. Click the Add Elevations icon, enter **CI ducts** for Name, and then enter **8'-8"** for Elevation.

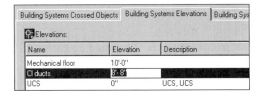

## Set the Catalog Paths and Building Systems Tooltips

10. Verify the catalog settings:

    a. Click the Building Systems Catalogs tab and verify the following catalog paths:

    **MvParts** Catalog is set to …\Program Files\Autodesk Building Systems 3\Aecb Catalogs 3\MvParts US Imperial\All Installed MvParts US Imperial.apc.

    **HVAC** Catalog is set to …\Program Files\Autodesk Building Systems 3\AECB Catalogs 3\Duct US Imperial\Duct US Imperial.apc.

**Piping** Catalog is set to ...\\*Program Files\Autodesk Building Systems 3\AECB Catalogs 3\Pipe US Imperial\Pipe US Imperial.apc*.

b.  Click the ellipses [...] button if you would like to see the drawings available.

11. Set the Tooltips:

a.  Click the Building Systems Tooltips tab.

b.  Verify that the Object Name, System, and Elevation are checked.

c.  Click OK.

The Catalog Path names may be set to a networked location for your office. The settings made here are used in the remaining exercises in this book. If your office has set these differently, remember to change them back once you are done with the exercises.

When activated, the Building Systems Tooltips allow you to hover the cursor over a Building Systems object to see any of this information about the object. Hovering the cursor over a duct enables you to see the duct's elevation and system with the tooltip.

You now know how to specify the Options for a mechanical drawing. Most of these settings are stored in the drawing. However, some of the background variables are also stored in the registry. If these settings are found in a drawing, then Building Systems will use the settings from the drawing. However, if you open up a drawing that has not been opened in Building Systems before, the drawing will not contain any of these settings, so Building Systems will read the values out of the registry and store them in the drawing the first time you save it using the Building Systems software.

The Building Systems Options are usually set at the beginning of a project and ignored until the next project comes around. In addition to the options, there are Layout Preferences for both ducts and pipes. The Layout Preferences store several key values for how ducts will be placed in the drawing. The next exercise takes you through setting the duct layout preferences. Layout preferences exist for the pipes also, but we will not cover those here.

## EXERCISE 2: SPECIFYING THE DUCT LAYOUT PREFERENCES

The Duct Layout Preferences dialog box has five tabs. Each of these tabs effects how ducts will be added to the drawing. Many of the settings on the tabs can be accessed directly from the Add Ducts dialog box.

**Open:** *04 Layout Preferences.dwg*

**Review the Duct Layout Preferences**

1. From the Mechanical menu, click Ducts ➤ Preferences.

2. Specify the Routing:

   a. Click the Routing tab.

   b. Verify the routing preferences are as shown in the following illustration.

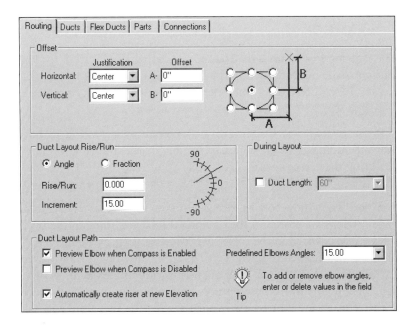

Of the four areas in the routing tab, Offset and Duct Layout Rise/Run are also available in the Add Duct dialog box while you are adding ducts. The During Layout area allows you to enter a value to break long stretches of duct you add it. The Duct Layout Path area includes a critical setting—the "Automatically create riser at new Elevation" toggle. This toggle determines if a vertical segment of duct is added when you specify a new elevation, or if the software waits for another pick point and adds a sloped transition to the new elevation.

3. Set the Duct Variables:

   a. Click the Ducts tab.

   b. Leave all of these values unchecked.

   The Lining and Insulation values toggle on and off adding insulation or lining automatically when you draw a duct. The value set here also applies to the fittings, since they are created at the same time you create a turn in the duct run. It is much easier to have the software automatically add the insulation to the fittings than to manually add this later.

4.  Specify Flex Duct Preferences:

    a.  Click the Flex Ducts tab.

    b.  Use the drop-down menu to change the 1-line annotation to Curve Pattern.

    c.  Use the drop-down menu to change the 2-line annotation to Vertical Pattern.

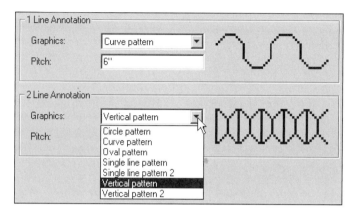

5.  Specify the default fittings and connection type:

    a.  Click the Parts tab.

    b.  Verify the connection type is set to slip joint.

    c.  Verify there is a part selected for each of the types of fittings shown in the list.

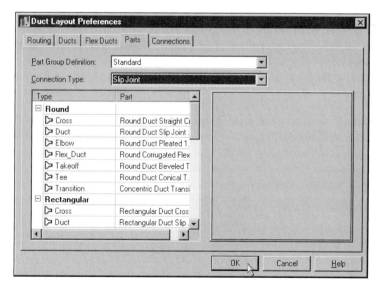

The Connection Type is set here, and every duct you place in the drawing will have this type of connector. The List below this is of the default fittings that will be placed in the drawing automatically. You can set the defaults here by clicking on the part name to activate a drop-down list of the available fittings for that condition. The Part Group Definition is for storing groups of default fittings. If you want to use a Part Group Definition, you must first create one using the Style Manager. Once a part group definition is created, you can use the drop-down to set it here. As changes are made to the list of default parts, they will be remembered by the part group definition that is active at the time.

6. Specify Connections:

    a.   Click the Connections tab.

    b.   Verify that the Connect with Duct is set to Use Tee.

    c.   Verify that the Terminal-Duct connection is set to Flexible.

    d.   Click OK.

The values you set here in the Duct Layout Preferences dialog box will affect many things about how you add ducts to the drawing. You will become very familiar with this dialog box because as you add ducts you will need to open it and change these values frequently. When you are adding ducts, you do not have to use the menu to access this dialog box—there is a direct link from the Add Duct dialog.

## LESSON 2: PLACING THE EQUIPMENT

When you start drafting a mechanical plan in AutoCAD, you might start by importing the architectural background and then adding blocks to represent the equipment or lines to represent the ducts. In AutoCAD, it does not matter which you add first because the lines and blocks do not have any dynamic relationship to each other. When you use Building Systems, however, you will add the MvParts that represent the equipment first. The reasoning behind this is because the MvParts will be used as the start and endpoints of the ducts or pipes that you will add later.

This lesson contains the following exercises:

- Adding a VAV Unit Using the Compass
- Building a Modular Air-Handling Unit
- Adding Return Air Grilles Using Preset Elevations
- Adding Air Diffusers

Each exercise has a drawing file associated with it. When you complete the exercises in this lesson, you will have added several types of MvParts to the drawing. You will also have utilized several of the tools that Building Systems provides to facilitate working in 3D. To complete these exercises you will be working with existing drawings. At the end of each exercise, there is a screen shot of what your finished drawing should look like.

### EXERCISE 1: ADDING A VAV UNIT USING THE COMPASS

The compass gives you the ability to guide the angle of your duct or pipe run using snap increments and plane rotations. In this lesson, you change the compass settings and then use the compass to insert an MvPart at a specific rotation angle. You will work more with the compass and ducts in the next lesson. The MvParts connectors may be used to place the part as well as the starting point for ducts or pipes.

 **Open:** *04 Using the Compass.dwg*

### Set the Compass and Snap Settings

1.  Change the Compass settings:

    a.  From the MEP Common menu, click Compass ➤ Compass Settings.

    b.  Change the color to blue so that you can see it on a white background. (Leave the compass color yellow if your background is black.)

    c.  Change the snap and tick mark increments to **45** degrees and click OK.

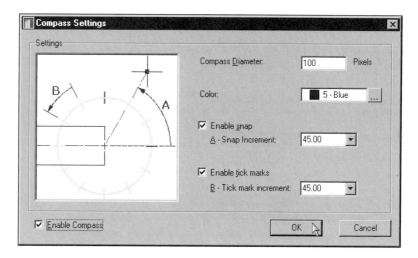

2.  Turn on the Node OSNAP:

    a.  Right-click on the OSNAPs button at the bottom of the screen and click Settings.

    b.  Select Object Snap On and select Node.

    c.  Make sure that no other snap settings are selected.

    d.  Click the Building Systems Snaps tab.

e.  Verify all the Building Systems snaps are selected.

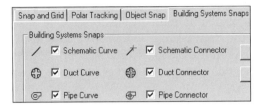

f.  Click OK.

## Insert a VAV Box MvPart

3.  Navigate to a Parallel Fan Powered VAV box:

a.  From the MEP Common menu, click MvParts ➤ Add.

b.  Click the Part tab, and, in the left panel, use the tree to browse to VAV Units ➤ VAV boxes ➤ Parallel Fan Powered VAV.

c.  For Part Size Name, select the 10 Inch Parallel Fan Powered VAV.

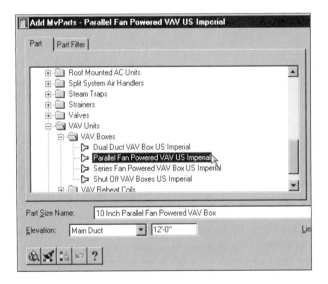

**Note:** There is a thumbtack in the upper-right corner of the Add MvPart dialog box. This thumbtack will affect the steps it takes to place a part in the drawing. If the thumbtack is "in" then the dialog box remains open at all times. In this case, it takes one mouse-click to move the focus of the cursor out of the dialog box and into the drawing. It takes another click to establish the insertion point of the MvPart or duct. If the thumbtack is "out," then the dialog box will minimize itself and automatically change the focus of the cursor to the drawing. In this case it only takes one mouse-click to establish the insertion point. The exercises in this chapter assume that the thumbtack is "out" while you are working.

4.  Specify the insertion properties for the VAV box:

    a.   Change the Elevation to Main Duct.

    b.   Verify the thumbtack at the upper-right corner of the dialog box is out.

    c.   Move the cursor into the left viewport.

    d.   At the command line, enter **B** three times to specify that the insertion basepoint be the connector of the round intake of the VAV.

    The **B** at the command line toggles the insertion point of the MvPart from the insertion point of the block through all the connectors on the MvPart.

5.  Insert the VAV box into your drawing:

    a.   Use the Node OSNAP to select the pre-drawn round node in the drawing.

    b.   Move the cursor around at the rotation prompt until you see 180 degrees.

    The compass only snaps to the increment defined in the Compass Settings dialog box; in this case, every 45 degrees.

    c.   Pick the point at 180 degrees to place the VAV box, and click Close on the Add MvPart dialog box.

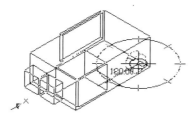

## Add Ducts Using the Compass

6.  Add a duct to the VAV box:

    a.   From the Mechanical menu, click Duct ➤ Add.

    b.   Verify that the thumbtack on the Add Duct dialog box is out.

    c.   Click on the Duct Connector on the upper-left side of the VAV box for the duct starting point.

    d.   Draw a duct toward the upper-left about 4', click to end the segment, but do not press ENTER.

    If the thumbtack is in you can see the duct length as you drag it by watching the Length option on the Add Duct dialog box.

7. Change the compass drawing plane while adding a duct:

   a. Enter **P** at the command line to change the plane of the compass to be perpendicular to the duct.

   b. Drag the cursor upwards to draw a vertical segment of duct and click to end the segment.

   c. Enter **P** at the command line to change the drawing plane back to horizontal and continue the duct toward the upper-left of the drawing.

   d. Press ENTER to end the command.

 **Note:** Using the compass, you can adjust the angle of the system in plan mode, or you can use the compass as a 3D drawing aid.

This is what your finished drawing should look like.

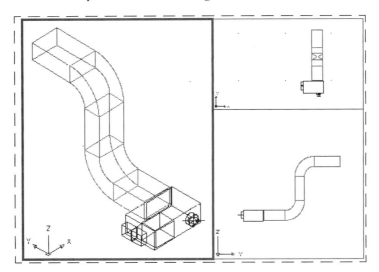

You have completed the task of using the compass to insert an MvPart VAV box, as well as drawn ductwork from the connector on the VAV unit. This exercise illustrates several things about connectors. MvParts can have multiple connectors. You can use any connector on the part as the insertion point of the part. When you add a duct to a connector, the duct will automatically adjust its size to the size stored in the connector. In addition, you saw that although Building Systems uses many dialog boxes, you should pay attention to the command line as well. Often there are options at the command line that may not be available in the dialog box.

In the remaining exercises in this lesson, you will build a modular air-handling unit using a combination of different MvParts, place equipment at particular elevations, and attach air diffusers to a ceiling grid.

### *EXERCISE 2: BUILDING A MODULAR AIR-HANDLING UNIT*

In this exercise, you will use different types of MvParts to build a modular air-handling unit. During this process, you use the compass to place the parts at specific rotation angles. You will use the connectors on the parts to place them relative to each other. In the drawing for this exercise, you will be working in the mechanical room on the second floor of the Challenger Learning Center.

**Open:** *04 Air Handling Unit.dwg*

### Start the Air-Handling Unit With a Mixing Box

1. Browse to an MvPart AHU Mixing Unit:

    a. From the MEP Common menu, click MvParts ➤ Add.

    b. Click the Parts tab, and in the left panel, use the tree to browse to Modular Air-Handling Units ➤ AHU Mixing Boxes ➤ Mixing Boxes US Imperial.

    c. For Part Size Name, select the AHU Mixing Box for Part Size 40.

    d. Notice the blue arrows show the flow direction in the preview picture of the part.

2. Place the AHU Mixing Unit:

    a. At the command line, enter **73', 66'** for the insertion point.

    b. Drag the cursor to the left and use the compass to place the AHU Mixing Box at 180 degrees, and then press ENTER to end the command.

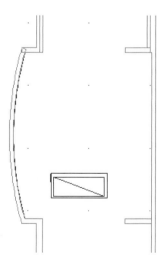

d.   Now that the insertion point is at the top edge of the coil, use the duct connector on the upper (north) edge of the AHU Mixing Unit to place the AHU Filter.

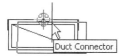

e.   Use the compass to place the part at 180 degrees rotation.

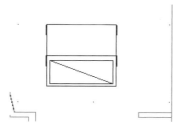

Using the connector on the filter as the basepoint and the connector on the intake as the insertion point allows you to place the pieces of equipment in line with each other.

You can continue to place parts without exiting the Add MvPart dialog box.

**Add a Coil to the Air Handler**

5.   Browse to AHU Coils:

a.   In the left panel, use the tree to go a few folders up, find AHU Coils, and select Large Coils US Imperial.

b.   Click the Details button.

Clicking the Details button enables you to see the variety of the parts of this type available in the catalog. You can also view connector information, layer key, and catalog ID information.

c.   Select the 40 SqFt Large AHU Coil Module Size 40.

6.   Place the AHU Coils:

a.   Click in the drawing to make it active and enter **B** at the command line to change the basepoint.

The flow through this part is in the positive Y direction, so the insertion point needs to be located on the bottom of the AHU Filter in plan view.

b.   Use the duct connector snap on the AHU Filter to place the AHU Coil above the AHU Filter with a 0 degree rotation.

**Add a Fan to the Air Handler**

7.  Browse to an AHU Fan:

    a.  In the left panel, use the tree to browse to AHU Fans. Select Fan Module Side Discharge US Imperial.

    b.  For Part Size Name, select the Fan Module Side Discharge for Part Size 40.

8.  Place the AHU Fan:

    a.  Click in the drawing to make it active and enter **B** twice at the command line to change the basepoint.

    b.  Use the duct connector snap on the AHU Coil to place the AHU Fan above the AHU Coil.

    c.  At the command line, enter **180** to specify the rotation of the part, and then press ENTER to end the command.

This is what your finished drawing should look like.

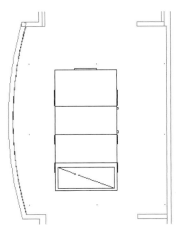

In this exercise, you built a size 40 modular air-handling unit by connecting a mixing unit, air filter, a coil, and a fan. In the next exercise, you will be adding return air grilles to your drawing using preset elevation values.

### EXERCISE 3: ADDING RETURN AIR GRILLES USING PRESET ELEVATIONS

One of the tools for creating a 3D model while working in the plan view is the Elevation setting. In this exercise you will be creating elevation presets to represent a 10' and an 8' ceiling. You then assign these elevations to return air grilles as you add them to your drawing.

 **Open:** *04 Air Return Grilles.dwg*

## Add Elevations in the Options Dialog Box.

1.  Specify the default elevation:

    a.  From the Tools Menu, click Options.

    b.  Click the Building Systems Elevations tab.

    c.  Click the Add Elevation icon on the lower left.

2.  Add an 8' ceiling elevation:

    a.  Enter **8' Ceiling** for name.

    b.  Enter **8'** for the height.

    c.  Enter **Lobby Ceiling** for Description.

3.  Add a 10' ceiling elevation:

    a.  Enter **10' Ceiling** for name.

    b.  Enter **10'** for the height.

    c.  Enter **Classroom Ceiling** for Description.

4.  Add a Main Trunk elevation:

    a.  Enter **Main Trunk** for name.

    b.  Enter **15'** for the height.

    c.  Enter **Main Duct Height** for Description, click Apply, and then click OK.

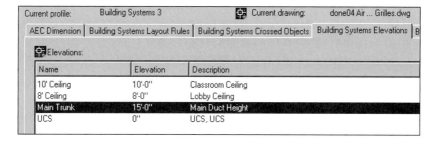

## Select and Add a Return Air Grille

5.  Specify a return air grille at an 8' ceiling elevation:

    a.  From the Mechanical menu, click Add Mechanical Equipment ➤ Air Terminal.

        This menu also accesses the Add MvParts function, but filters the MvParts list to only air terminals.

    b.  Click the Part tab, and in the left panel browse to Grilles ➤ Ceiling Return Air Grilles with Trim US Imperial.

 c. Select 10 inch x 6 inch Return Air Grille with Trim for Part Size Name.

 d. Select 8' ceiling for Elevation.

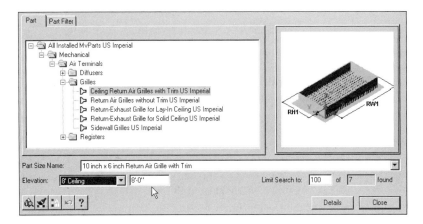

6. Add the Return Air Grille:

 a. Click in the drawing to make it active.

 b. Click in the electrical room 135, located at the upper-right portion of the drawing, to add the return air grill.

 c. Use the compass to place the grille at 0 rotation.

7. Add a second return air grille at an 8' ceiling elevation:

 a. In the Add MvParts dialog box, change the Part Size Name to 24 inch x 12 inch Return Air Grille with Trim.

 b. Click in the drawing to make it active.

 c. Click in Virtual Center room 134, and use the compass to place the grill at 0 rotation.

## Add Classroom Return Grilles

8. Specify a return air grille at a 10' ceiling elevation:

 a. In the left panel of the Add MvParts dialog box, browse to Return Air Grilles without Trim US Imperial.

 b. Select 24 inch x 24 inch Return Air Grille for Part Size Name.

 c. Select 10' ceiling for Elevation.

9. Add the return air grille to classroom 154:

 a. Click in the drawing to make it active.

 b. Click in the lower right-hand corner of classroom 154. At the command line enter **0** for rotation and then press ENTER.

10. Add a second return air grille to classroom 154:

    a.  In the Add MvParts dialog box, select 24 inch x 12 inch Return Air Grille for Part Size Name.

    b.  Click in the drawing to make it active.

    c.  Click in classroom 154 just above the existing return air grille. At the command line enter 0 for rotation and then press ENTER.

11. Repeat steps 8–10 to add two more return air grilles to classroom 155.

This is what your finished drawing should look like.

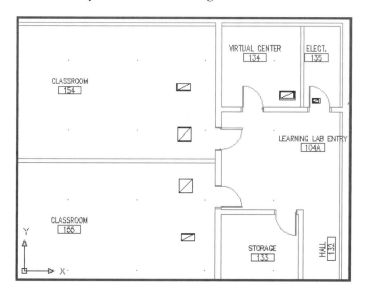

In this exercise, you created elevations and assigned these elevations to the return grilles as you added them to the drawing. In the next exercise you will add a ceiling grid to a drawing and use the ceiling grid to place air diffusers.

## EXERCISE 4: ADDING AIR DIFFUSERS

A ceiling grid may or may not exist in the drawings you receive from the architect. In this drawing, one classroom already contains a ceiling grid that is from an externally referenced (xrefed) background drawing. This ceiling grid is an Architectural Desktop Ceiling grid object, not lines. In this exercise you will create another ceiling grid object for the other classroom. You then use intersections and nodes to place air diffusers in the drawing.

While MvParts may be added at a preset elevation, some actions override the elevation setting in the Add MvParts dialog box. If you add an MvPart using a standard AutoCAD OSNAP, the part will be placed at the height (or Z value) found by the OSNAP. This is true of otrack as well. When you use otrack to place an MvPart or start a duct, the insertion height will be determined by the Z value of the original tracking point.

 **Open:** *04 Air Diffusers.dwg*

### Add a Polyline for the Ceiling Grid Boundary and Add a Ceiling Grid

1. Set up your drawing to create a ceiling grid:

   a. Use endpoint OSNAP to draw a rectangle on the inside walls of classroom 154.

   The rectangular polyline will be used later to clip the ceiling grid. You can also use Architectural Desktop space objects to define the ceiling grid. When a space object is used as a ceiling grid boundary, the ceiling grid is inserted at the elevation specified for the ceiling of the space object. Because you will use a polyline for the ceiling grid boundary, the grid will be created at the elevation of the polyline.

   b. From the View menu, click 3D Views ➤ NE Isometric.

2. Create a ceiling grid:

   a. From the MEP Common menu, click Grids ➤ Add Ceiling Grid.

   b. Enter **40'** for X-Width, and verify that 20' is entered for Y-Depth.

   c. Enter **2'** for the X and Y Baysize.

3. Add the ceiling grid:

   a. Click Set Boundary and select the rectangle you created in step 1 as the boundary for your ceiling grid.

   b. Select the upper corner of the classroom as the insertion point and press ENTER to accept 0 as the rotation angle.

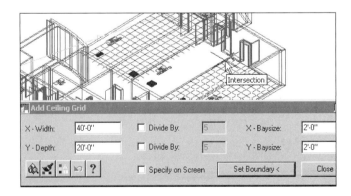

   c. Press ENTER to end the command.

   The ceiling grid is inserted at a 0 elevation.

4.  Change the ceiling grid elevation:

    a.  Select the ceiling grid, right-click, and select Ceiling Grid Properties.

    b.  Click the Location tab and enter **10'** for Z value.

    c.  Click OK.

    The elevation of the ceiling grid automatically updates to the correct ceiling height of 10'.

**Add Ceiling Diffusers**

5.  Change the drawing views:

    a.  From the Views menu, click 3D Views ➤ Top.

    Notice that the ceiling grid disappears. When working with ceiling grids, by default, the ceiling grid is only visible in isometric views, or in one of the RCP (Reflected Ceiling Grid) layouts. This is controlled by the Display Set, accessible through the Display Manager.

    b.  Click the Plan 2 Line–RCP tab at the bottom of the drawing.

    You should now be able to see both of the ceiling grids.

6.  Browse to the Ceiling Diffuser:

    a.  Specify that your OSNAPs be set to Node only.

    b.  From the Mechanical menu, click Add Mechanical Equipment ➤ Air Terminal.

    c.  Click the Part tab and browse to Perforated Face Round Neck Ceiling Diffusers US Imperial.

7.  Add Ceiling Diffusers:

    a.  For Part Size Name select Perforated Face Neck Ceiling Diffusers – 10 Inch Neck – 24 x 24 Face.

    b.  Move the cursor over the ceiling grid in classroom 154.

    You will see that there are nodes at each intersection.

    c.  Place five ceiling diffusers as shown.

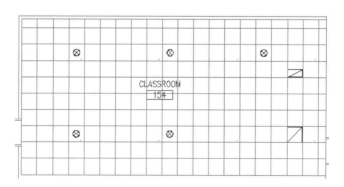

Ceiling Grid objects within xrefs (such as the one in Classroom 155) do not give access to nodes. However, you can use the Intersection OSNAP when working in an isometric view, or change the representation in the current display set from background representation to the model representation.

8. Add ceiling diffusers to the externally referenced ceiling grid:

   a. Click the Model tab and zoom into classroom 155.

   b. From the View menu, click 3D Views ➤ Northeast Isometric.

   c. Using the intersection OSNAP, place five ceiling diffusers as shown.

This is what your finished drawing should look like.

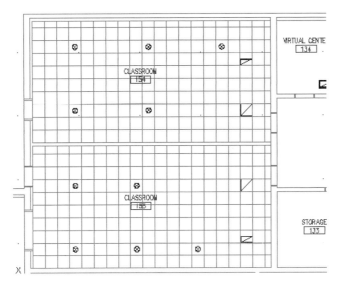

In this exercise, you added an Architectural Desktop ceiling grid and then attached air diffusers to the ceiling grid.

In Lesson 2, "Placing the Equipment," you completed four exercises dealing with adding MvParts. You have seen that objects are very dialog box–intensive. In the next lesson you will draw ducts to connect the pieces of equipment you placed in the last exercise. When you add ducts, several similarities will be evident. When adding the ducts, you will have access to the compass as you did in "Adding a VAV Unit Using the Compass" and "Building a Modular Air-Handling Unit." The Elevation variable you used in "Adding Return Air Grilles using Preset Elevations" will be available as well when you add ducts. Ducts are added to the connectors of the MvParts in a similar fashion to the way you located the coil on the main air-handling unit using the Building Systems Connection snap points.

## LESSON 3: WORKING WITH DUCTS AND FITTINGS

Now that you have placed the equipment in the drawing, it is time to connect them with ducts. When you add ducts, you will be working with the compass and the connectors on the MvParts. This lesson is divided into three groups of exercises.

The first three exercises in this lesson focus on controlling the fittings that are added automatically by the software. The control for the fitting is stored in the duct layout preferences. You can access the duct layout preferences directly by picking the Preferences icon in the Add Duct dialog box. Additionally, on the add duct dialog box, there is an arrow next to the layout preferences icon that allows you to access the default fittings without opening the duct layout preferences.

You have seen the Building Systems *duct connector* on the MvParts, which aides in attaching ducts to equipment. On ducts, there are two types of connectors. In addition to a *duct connector* at the end of the duct, a *duct curve connector* exists along the centerline of every duct. If you add a new segment of duct to an existing segment the type of connector you attach to determines the result. If you start the new segment at the duct connector at the end of the duct, the new duct length will be added to the old. If the new duct starts with a duct curve connector, the software assumes you are creating a branch and a tee or tap will be added to the existing duct. When you are adding ducts to the end of an existing duct, make sure you are connecting to the duct connector, not the duct curve connector.

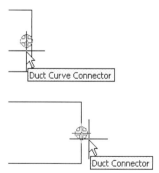

The second group of three exercises shows you different methods for routing the duct through the building model. Building Systems provides several different ways to route ducts through the model. Auto-routing allows you to route ducts both in plan and 3D. For situations that auto-routing will not accommodate, you can manually route ducts around obstacles using the compass. In the final exercise you will add a flex duct to the drawing.

This lesson contains the following exercises:

- Drawing Ductwork
- Changing Duct Fitting Preferences as You Draft
- Connecting Ductwork With Tees and Taps
- Routing Ducts Around Beams Using the Compass
- Routing Ducts Using Elevation Changes
- Auto Routing Ductwork
- Adding Flex Duct

Each exercise has a drawing file associated with it. To complete these exercises you will be working with existing drawings. At the end of each exercise, there is a screen shot of what your finished drawing should look like.

### EXERCISE 1: DRAWING DUCTWORK

In this exercise, you will be adding ductwork to represent the main supply duct for part of the building. During this process, you will be changing the size and the direction of the duct run. To complete this exercise you should turn off OSNAPS, Polar Tracking, and Otrack.

**Open:** *04 Ducts 2D.dwg*

**Add a Duct to the Drawing**

1. Set the Duct Variables:

   a. From the Mechanical menu, click Ducts ➤ Add.

   b. Change the System to Supply, the Shape to Round, and select 20" for Diameter.

   c. Verify that the Horizontal and Vertical Justification is Center and the Offset is 0.

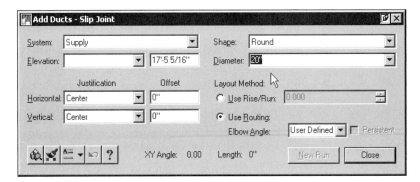

2. Start the duct run:

   a. Move the cursor into the drawing and use the node Osnap to select the red point that is in the lower-right corner of the drawing for the starting point of the duct.

   b. Drag the duct to the left.

      The compass constrains the duct to 45 degree increments (set in exercise 1 of lesson 2, earlier this chapter).

   c. At the command line, enter **7'**.

3. Change the direction of the run to add an elbow:

   a. Drag the cursor upwards and enter **12'** at the command line.

      Whichever round elbow is specified as the default in the duct layout preferences is added to the duct as a fitting.

   b. In the Add Ducts dialog box, change the Diameter to 14".

   c. Drag the cursor upwards and pick a point beyond the upper VAV box.

   d. Press ENTER to end the command.

This is what your finished drawing should look like.

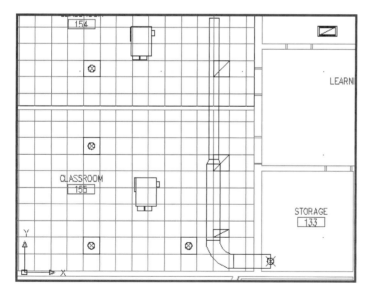

In this exercise you added the main supply ductwork for part of the building. As you manually routed the ducts through the building, the fittings were automatically placed for you. As you turned the corner with the duct, the elbow was added to the run. As you changed the size of the duct, a concentric reducer was added to the run. The type of fitting placed was controlled by the duct layout preferences. In the next exercise you access these settings as you are adding the duct.

*EXERCISE 2: CHANGING DUCT FITTING PREFERENCES AS YOU DRAFT*

In this exercise, you will use the justification and offset preferences in the Add Ducts dialog box, as well as learn how to change your duct fitting preferences as you draft. Before you start this exercise, open the snap settings and turn off all the AutoCAD object snaps. Leave on all the Building Systems snaps.

 **Open:** *04 Duct Preferences on the Fly.dwg*

### Start the Duct

1. Specify the duct preferences:

   a. From the Mechanical menu, click Duct ➤ Add.

   b. Select Supply for the system.

   c. Change the shape to Rectangular.

   d. Click the down arrow next to the Preferences icon, browse to elbow, and select Rectangular Mitered Elbow.

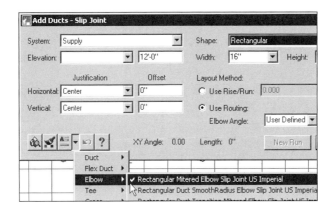

### Add a Duct With an Elbow

2. Draw the first segment of the duct:

   a. Move the cursor into the drawing.

   b. Select the duct connector on the upper-left side of the VAV box to start the duct run.

   c. Drag the duct to the left and click an endpoint for this segment of duct near the left wall of Classroom 155.

   The connector on the VAV box contains size and shape information that is read into the Add Duct dialog box. Notice that the duct shape automatically changes to rectangular, the duct width changes to 20", and the duct height changes to 12".

3.  Draw an offset segment of the duct.

    a.  In the Add Duct dialog box, change the Horizontal Justification to Right and the Offset to 6".

    b.  Drag the cursor down and pick a point just above the southern wall to classroom 155.

        A rectangular mitered elbow fitting was placed at the location where you changed the direction of your duct run. Additionally, the offset you specified placed the duct 6" from its start and endpoints. Offset only works when the duct is added to the drawing. The offset setting is not stored in the duct itself.

## Add an Eccentric Reducer

4.  Reduce the duct size:

    a.  In the Add Duct dialog box, change the Width to 14".

    b.  Click the down arrow next to the Preferences icon, browse to Transition, and select Rectangular Duct Eccentric Transition.

        You need to change the reducer to eccentric in order to keep the duct justified to the side of the wall.

    c.  Click once in the drawing to make it active and pick a point just below the wall of classroom 155.

    d.  Click Close in the Add Duct dialog box.

This is what your finished drawing should look like.

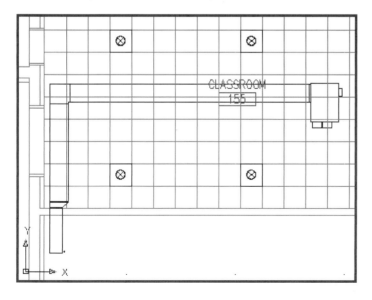

In this exercise you used the justification and offset preferences in the Add Ducts dialog box to draw a duct offset 6 inches from the right hand wall. You also changed the fitting preferences on the fly directly from the Add Duct dialog box.

## EXERCISE 3: CONNECTING DUCTWORK WITH TEES AND TAPS

In this exercise you will add branch ducts. When connecting a duct to another duct, either a tee or a tap is automatically placed in the drawing. You can control whether a tee or a tap is inserted into your drawing using Duct Preferences.

Although you specify a tee or tap in the connections tab of the preferences, you still have control over what type of tee or tap is inserted by using the Parts tab. The following exercise starts with the part preferences for a tee as the rectangular tee. When you add the branch duct and specify a smaller size, you will be presented with a warning message that the regular tee will not work in this situation. You will be asked to choose a transitional tee to accommodate the different size branch.

**Open:** _04 t and tap.dwg_

### Add a Branch and Tee

1. Specify a duct connection preference:

   a. From the Mechanical menu, click Ducts ➤ Duct Preferences.

   b. Click the Connections tab.

   c. Select Use Tee.

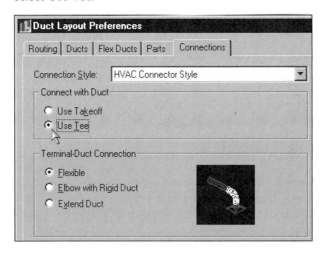

2.  Specify a takeoff part:

    a.  Click the Parts tab.

    b.  Change the Rectangular ➤ Takeoff to Rectangular Duct Plain Tap.

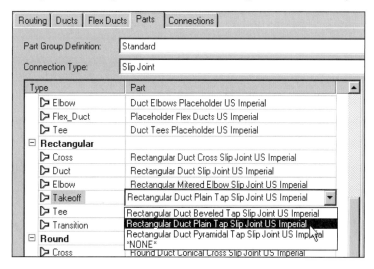

    c.  Click OK in the preferences dialog box.

3.  Add branch duct with transitional tee:

    a.  From the Mechanical menu, click Duct ➤ Add.

    b.  Use the Duct Curve Connector snap to pick a startpoint in between the double doors.

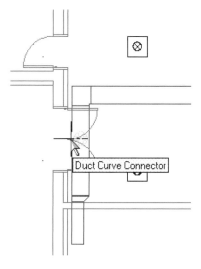

c.   In the Add Duct dialog box, change the Width to 14".

d.   A warning message appears; click Choose a Part.

e.   In the Choose a Part dialog box, select Regular Duct Transition Tee Slip Joint and then click OK to return to the Add Duct dialog box.

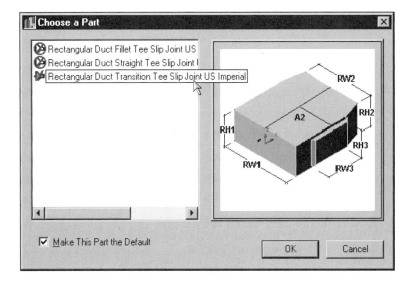

f.   Change the Height to 10".

4.   Continue to add the duct:

a.   Drag the duct to the left and pick a point.

b.   Pick New Run in the Add Ducts dialog box.

The duct segment is completed and the transitional tee is automatically inserted.

## Add a Branch and Takeoff

5.   Change the Duct Preferences:

a.   Click the Properties icon in the Add Ducts dialog box.

b.   Click the Connections tab.

c.   Select Use Takeoff and click OK.

6.  Add more duct:

    a.   From the Mechanical menu, click Duct ➤ Add.

    b.   Pick a point just below the wall of classroom 155.

    c.   Drag the cursor to the right and pick a point to end the segment.

        A tap duct fitting is placed instead of a tee.

    d.   Click OK in the Add Ducts Dialog box.

This is what your finished drawing should look like.

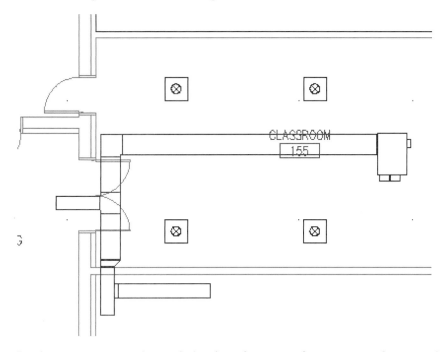

In this exercise you changed the duct fitting preferences in order to make the default duct fitting either a tee or a tap. You have used the automatic fitting insertion routines that are part of adding a duct.

You can also add fittings manually. Offsets, tees, and wyes are just a few of the types of fittings that are not part of the automatic placement routine. You add these fittings in the same manner as MvParts, selecting a size from the catalog and placing it in the drawing. When you add fittings manually, use the parts filter tab to isolate the part you need. In practice, it is easier to place the fitting in the drawing and then move it into place than to attempt to place the fitting directly on the duct. Use the Building Systems Snaps tool bar to aid in moving the fitting onto the duct.

## *EXERCISE 4: ROUTING DUCTS AROUND BEAMS USING THE COMPASS*

In this exercise, you will use the compass to change the rotation angle of the duct run in order to route the ductwork around a beam.

**Open:** *04 Routing with Compass.dwg*

### Add a Duct

1.  Prepare your drawing environment:

    a.  Verify that ortho, otrack, and polar are all turned off.

    b.  Click in the upper-right viewport to make it current.

    c.  From the Mechanical menu, click Duct ➤ Add.

2.  Specify the duct properties:

    a.  Select Supply for System and select Rectangular for Shape.

    b.  For Width select 30" and for Height select 24".

    c.  Verify that Horizontal and Vertical Justification is Center and the Offset is 0.

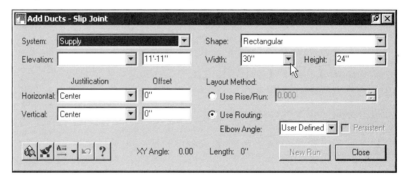

3.  Add ductwork:

    a.  Click in the upper-right viewport to make the drawing active, and pick a point at the middle left to specify the start of the duct run.

    b.  Drag the cursor to the right about 16' and click to end the segment.

    You can monitor the length in the Add Ducts dialog box. The length is dynamically updated as you drag your cursor.

    c.  Click once in the lower right-hand viewport to make it active.

    Notice that the compass is active in the viewport, but appears as a line because you are looking at it from the side. You may see the compass in this orientation at other times. For example, if you add a part in a plan that breaks into a duct such as a damper, you get the compass because you are adding a part. But because the part is being added perpendicular to the duct, the compass will appear on edge, or as a line in the drawing.

### Route the Duct Under the Beam

4.  Change the drawing plane:

    a.  Enter **P** twice at the command line to change the plane of the compass and to swing it into view.

    b.  Use the compass to pick 4 more points, routing the duct under the beam.

This is what your finished drawing should look like.

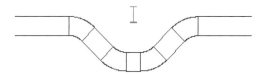

In this exercise, you used the compass to route a duct run around a beam by switching the compass plane, as well as using the compass to define the rotation angle of the duct run. While this exercise has several viewports set up to illustrate what happens as you modify the compass plane, you will probably find it easier to do this while looking at an isometric view.

### EXERCISE 5: ROUTING DUCTS USING ELEVATION CHANGES

In this exercise you will use routing and elbow angles to route the same duct as the last exercise.

**Open:** *04 Routing with elbows.dwg*

### Set the Variables for the Duct

1.  Prepare your drawing environment:

    a.  Verify that ortho, Otrack, and polar are all turned off.

    b.  Click in the upper-right viewport to make it current.

    c.  From the Mechanical menu, click Duct ➤ Add.

2.  Specify the duct properties:

    a.  Select Supply for System and select Rectangular for Shape.

    b.  For Width select 30", and for Height select 24".

    c.  Type **13'** in the Elevation line.

    d.  Verify that Horizontal and Vertical Justification is Center and the Offset is 0.

3. Set the Routing Preferences:

   a. Click the Preferences icon.

   b. On the Routing tab, at the lower left uncheck the Automatically create riser at new Elevation box.

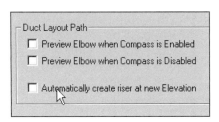

   c. Click OK to return to the Add Ducts dialog box.

## Add the Duct Under the Beam

4. Add the first section of duct:

   a. Move the cursor into the upper-right viewport.

   b. Pick a point at the middle left to specify the start of the duct run.

   c. Drag the cursor to the right about 14' and click to end the segment.

5. Add the duct under the beam:

   a. In the Add Ducts dialog box, type **11'** for the Elevation.

   b. Move the cursor back into the upper-right viewport.

   c. Click a point just to the left of the beam, in line with the first segment of duct.

   d. Click Yes to accept Custom Sizes.

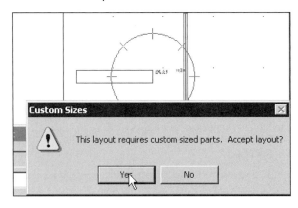

   e. Click a point past the beam on the right side.

   f. Click Yes to accept Custom Sizes.

   g. In the Add Ducts dialog box, type **13'** for the elevation.

h.   Click a point about 3' further to the right.

i.   Click Yes to accept Custom Sizes.

j.   Click a point at the right side of the viewport.

k.   Click Yes to accept Custom Sizes.

l.   Click OK to end the command.

This is what your finished drawing should look like.

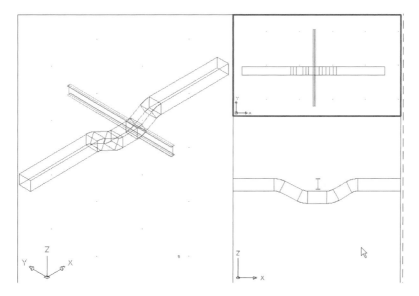

In this exercise, you routed the duct around the beam by changing elevations while drawing in plan view.

### EXERCISE 6: AUTO ROUTING DUCTWORK

In this exercise, you will use Building Systems' auto routing feature. You will use this feature both to connect two ducts, and to connect an air terminal to a branch duct. In order for auto routing from air terminals to work, the connector of the air terminal must be below the duct to which you are connecting. How the air terminal is connected to the duct is determined by using the Connections tab on the Duct Preferences dialog box.

**Open:** *04 AutoRouting.dwg*

**Use Auto Routing to Connect Two Ducts in 3D**

1.   Specify Routing Settings:

a.   Select the round duct at the number 1 in the drawing.

b.  Right-click, and click Add Selected.

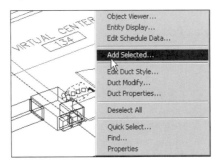

c.  In the Add Ducts dialog box, change the elbow angle to 90 degrees and check the Persistent box.

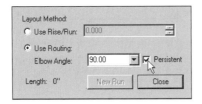

d.  Start the duct at the duct connector at the free end of the duct near number 1.

e.  Click the duct connector at the free end of the duct near number 2.

f.  Enter **C** at the command line to "connect" to the different elevation.

g.  Enter **N** two times at the command line to see the other auto routing options.

h.  Enter **A** at the command line to accept solution number four.

2.  Repeat, connecting ducts 3 and 4.

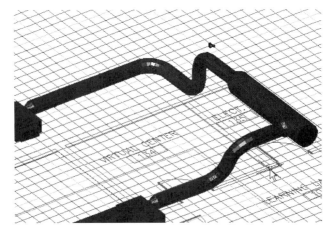

**Use Auto Route to Connect an Air Terminal to a Branch Duct**

3.   Click the Plan–2 Line layout tab.

4.   Add a duct:

    a.   From the Mechanical menu, click Duct ➤ Add.

    b.   Click the duct connector on the air terminal at point 5 in the drawing as the start point of the duct.

    c.   Click the duct curve connector at point 6 for the endpoint of the duct.

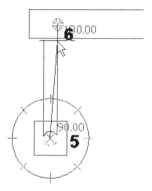

    d.   Press ENTER 3 times to connect to the new elevation, accept the solution offered, and to end the command.

This is what your finished drawing should look like.

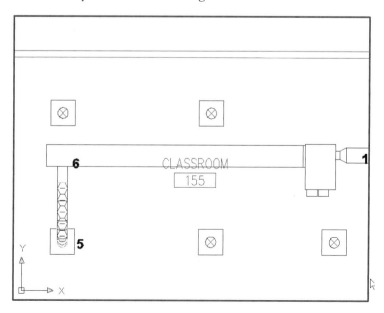

In this exercise you used Building Systems' auto routing in order to connect ducts to ducts and a terminal to a branch duct without elevations. Auto routing is a powerful tool and will also work when drawing ducts at the same elevation in plan view. Experiment with the user set elbow angles in the Add Ducts dialog box to add two 90 degree angles at once while drawing ducts in the plan.

### EXERCISE 7: ADDING FLEX DUCT

In this exercise you will add three pieces of flex duct. The first piece of flex duct you add connects the main duct to VAV boxes using auto routing. The VAV boxes in this drawing are suspended between two joists and are higher than the main duct. Therefore, the flex duct automatically makes a connection between the two elevations. You create the second piece of flex duct by selecting a polyline and converting it to a flex duct. You add the third piece of flex duct by manually connecting an air diffuser to the system. Before you start this exercise, change the snaps on the Building Systems compass to 15 degree increments.

**Open:** *04 Flex to VAV.dwg*

### Connect the End of Duct to the VAV unit with Flex Duct

1.  Add the first piece of flex duct:

    a.  From the Mechanical menu, click Flex Ducts ➤ Add Flex Duct.

    b.  Click in the left viewport, start the duct at the end of duct 1, and press ENTER to connect to the different elevations.

    c.  Pick the duct connector on the VAV box as the second point and press ENTER twice.

    The duct and the VAV are connected with flex duct, and the flex duct adjusts to the differing elevations. A transition is also inserted to accommodate the smaller diameter intake on the VAV box.

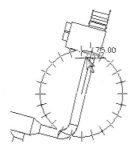

### Convert a Polyline to Flex Duct

2.  Convert a polyline to a flex duct:

    a.  From the Mechanical menu, click Flex Ducts ➤ Convert Flex Duct.

    b.  Click the polyline as the object to convert and press ENTER.

    c.  Enter **Y** at the command line to erase the polyline.

        The Modify Flex Duct dialog box appears.

3.  Set the variables of the flex duct:

    a.  Change the System to Supply.

    b.  Change the Diameter to 12".

    c.  Click OK.

        The flex duct is added to the drawing.

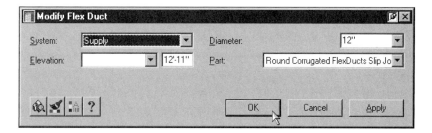

4.  Add a flex duct to an air terminal:

    a.  From the Mechanical menu, click Flex Ducts ➤ Add Flex Duct.

    b.  Pick the end of the small piece of duct that points towards the diffuser, and press ENTER.

    c.  Draw a small piece of flex duct in the direction that the duct is pointing (75 degrees on the compass).

    d.  Pick the connector on the air terminal and press ENTER twice to end the command.

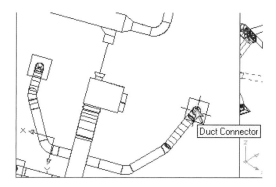

This is what your finished drawing should look like.

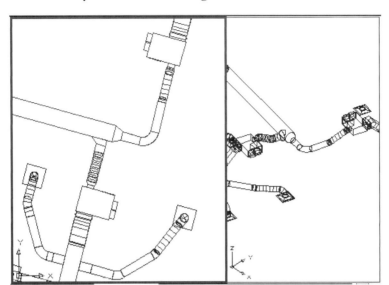

In this exercise you used flex duct to connect various pieces of the mechanical system. Here are a couple more tips on using flex duct in your drawings. When you are connecting flex to an air terminal, the process is backwards from that of using hard duct and auto-routing. Add a duct from the air terminal to the branch duct. Add flex duct from the branch duct to the air terminal. When you connect ducts and equipment that are very close together, use the spline segment mode when adding flex duct. The spline mode does a better job of contorting through tight turns than the straight or curve segment modes. Often, when you need a bit of flex duct between parts, it is easiest to connect the parts with regular duct using the auto-routing, then break out a bit of the duct and add flex duct between the existing duct and fitting.

In Lesson 3, "Working with Ducts and Fittings," you completed seven exercises demonstrating the controls of fittings and auto-routing the ducts. Using these tools, you should be able to tackle any duct routing task that you come across in your daily production.

## LESSON 4: MODIFYING YOUR MECHANICAL PLAN

As you work on a project, changes happen. The Building Systems objects have several features that are useful when it comes time to modify the building model.

This lesson contains the following exercises:

- Changing Duct Sizes
- Changing the Sizes of Parts

The first exercise illustrates the duct object's ability to translate a change in size down a main trunk, resizing all of the connected ducts at the same time. The second exercise targets the MvParts and how changing their size once they are in the drawing will affect the parts connected to them. Each exercise has a drawing file associated with it. At the end of each exercise, there is a screen shot of what your finished drawing should look like.

### EXERCISE 1: CHANGING DUCT SIZES

In this exercise you will resize a duct segment and a duct branch using the Modify Duct dialog box. The correct transition fittings are added to your drawing as you draft.

**Open:** *04 Resize Ducts.dwg*

### Resize a Single Duct Segment

1. Select the duct segment:

   a. From the Mechanical menu, click Duct ➤ Modify.

   b. In the left viewport, select the duct that sticks out to the right of the tee at the upper left-hand corner of the drawing.

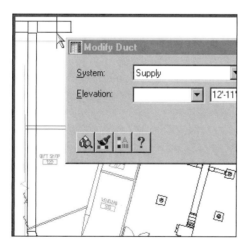

2. Change the size:

   a. In the Duct Modify dialog box, change the diameter to 14".

   b. Click OK.

      The duct is resized and a transition fitting is placed next to the tee fitting. If the tee fitting were a transitional tee, it would accommodate the new size duct.

3.  Resize a branch:

    a.  Select the section of duct that extends through Vending room 105, right-click, and click Duct Modify.

    b.  In the Duct Modify dialog box, change the diameter to 20" and click OK.

        The Transition Needed dialog box appears. The location in the drawing where the transition is needed is highlighted.

    c.  Click Re-Size Branch to a Junction/Transition and then click OK twice.

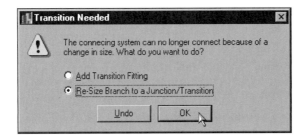

This is what your finished drawing should look like.

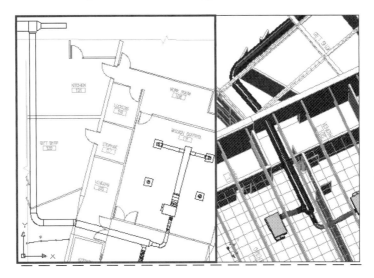

In this exercise you resized a duct segment and a duct branch. In the next exercise you will be resizing MvParts, and modifying the drawing using the AutoCAD MOVE command.

### EXERCISE 2: CHANGING THE SIZES OF PARTS

In this exercise, you will resize parts using the Modify MvPart dialog box. When you choose a new size for a part, the new part is pulled from the catalog and placed in your drawing. The new part is placed at the block insertion point, and, depending on the part, may need to be repositioned in the drawing. You can relocate parts using the AutoCAD nodes or the Building Systems connectors that are located at each connection point on the part. When you resize a part, the original size part and all of the parts associated with the blocks remain in the drawing. After resizing parts, the style definition of the older part size remains in the drawing and contains the definitions of all the AutoCAD blocks that make up the part. If you have made a lot of modifications to your drawing, you reduce the drawing size by using the BLDSYSPURGE command, which purges all unused MvPart definitions from the drawing. This command is available from the MEP Common menu as the utility Purge Building Systems Objects.

 **Open:** *04 Resize Parts.dwg*

**Turn on the Building Systems Snaps Toolbar**

1. From the Tools menu, select Customize ➤ Toolbars:

   a. Click the Menu Group BSCOMMON3.

   b. Click the checkbox for the Building Systems Snaps.

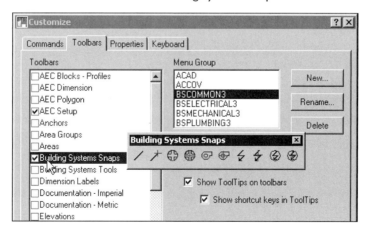

   c. Click Close in the Customize dialog box.

   d. Move the toolbar to an edge of the drawing to dock it.

### Change the Size of the VAV Box and Reheat Coil

2.  Change the size of the VAV box:

    a.  From the MEP Common menu, click MvParts ➤ Modify.

    b.  Select the VAV box.

    The MvPart Modify dialog box appears, showing that this MvPart is a 12 Inch Parallel Fan Powered VAV Box.

    c.  For Part Size, select 8 Inch Parallel Fan Powered VAV Box, and then click OK.

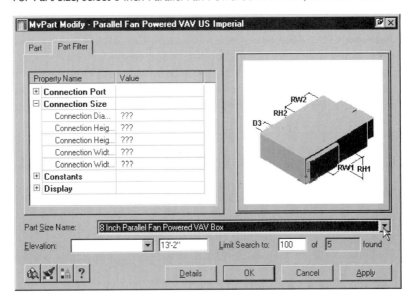

You may be prompted with a warning that tells you that the connection to the flex duct is being broken. If so, click OK.

The VAV box is resized based on the insertion point of the MvPart. In this case, the insertion point was at the lower left corner.

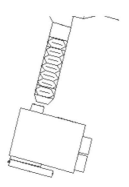

3.  Change the size of the reheat coil:

    a.  From the MEP Common menu, click MvParts ➤ Modify MvPart.

    b.  Select the reheat coil.

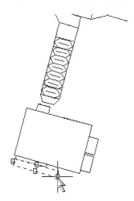

    The MvPart Modify dialog box appears, showing that this MvPart is a 28 x 16 in. Reheat Water Coil.

    c.  For Part Size, select 20 x 12 in. Reheat Water Coil to match the output of the 8" VAV box and then click OK.

**Move the Parts Back Together**

4.  Use the Building Systems snaps to move the VAV box:

    a.  Enter **M** at the command line.

    b.  Select the VAV box that is in plan view on the left and press ENTER.

    c.  Click the Duct Connector snap on the toolbar.

    d.  Click the supply intake on the VAV box as the basepoint.

    e.  Again, click the duct connector snap on the toolbar, then click the end of the fitting on the flex duct as the second point of displacement and press ENTER.

5.  Move the reheat coil to connect with the VAV box using the node OSNAP:

    a.  Enter **M** at the command line.

    b.  Select the reheat coil and press ENTER.

    c.  Use the node OSNAP to specify the connector on the inside face of the coil as the basepoint, and press ENTER.

d. Use the node OSNAP to specify the supply output on the VAV box as the point of displacement.

 **Note:** Building Systems snaps are only active when you are running a Building Systems command. Therefore, using the node that is built into the connector is very useful when you are using standard AutoCAD commands such as MOVE.

This is what your finished drawing should look like.

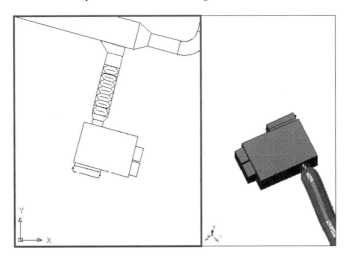

In this lesson, you used the tools to modify the building model. You can change duct sizes individually or for the entire run. When you change the size of MvParts, you will have to modify the connecting pieces as well.

## LESSON 5: WORKING WITH SYSTEM DEFINITIONS AND LAYER KEYS

System definitions are used for both schematic diagrams and the mechanical model. System definitions control layering, labeling information, and connectivity between the drawing elements of a schematic diagram. The system definition also provides display overrides for the sub objects in the system. When you create a schematic system definition or an HVAC system definition using the Style Manager, the variables you set in the Design Rules tab are the same. The HVAC system definition adds one more tab—the Rise and Drop tab—to give you control over the rise and drop symbols of a given system.

While very much alike in appearance and creation, the application of the system definitions is different for the schematic objects than for MvParts and ducts. Like the schematic system definition, you choose the system in the Add Ducts dialog box as you create ducts, and can modify the system assignment after the duct branch is drawn. While you are adding ducts, the fittings that are automatically placed assume the system of the duct you are drawing. Unlike schematic lines and symbols, the HVAC system is assigned only to the ducts and not the MvParts as they are added to the drawing. Exercise three illustrates how the system definitions are used by the MvParts.

System definitions control layering in the following way. In the Design Rules tab of the system definition is an area that allows you to apply either a layer key to all parts of the system, or an override to the ducts' and fittings' hard-coded layer keys. When ducts and fittings are drawn they refer to these layer settings in the system definition.

For example: A duct and fitting are drawn using the "Standard" system definition.

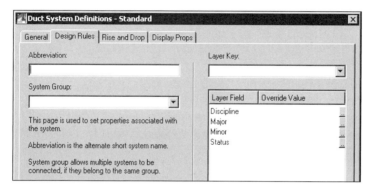

When the duct is added, it brings with it its hard-coded layer key *Duct*. In the layer key style the key *Duct* is assigned the layer M-STND-Duct. Likewise, the fitting brings with it its layer key *DuctFitting*, which it tied in the layer key style to the layer M-STND-Dfit. Within the Standard system definition, no layer keys or overrides are assigned, and so the duct will be placed on layer M-STND-Duct, the fitting on M-STND-Dfit.

In the Building Systems templates, the sample system definitions use overrides to the major field of the layer name. They look like this:

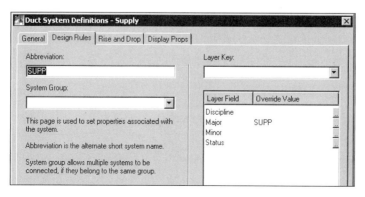

When a duct branch is drawn with this system definition, the duct and fitting look to the layer key style and find their layers M-STND-Duct and M-STND-Dfit. The system definition then replaces the second part (or the major field from the AIA layer naming convention) with SUPP. The resulting ducts end up on layer M-SUPP-Duct and fittings on layer M-SUPP-Dfit.

These two methods of defining system definitions will layer the duct and the fitting differently because the duct and the fitting have different layer keys. If you would like the ducts and fittings to be on the same layer, you can assign a layer key to the system definition. Most of the exercises in the book have been set up this way.

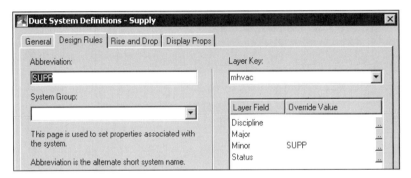

Setting a layer key in the system definition replaces the hard-coded layer keys of the duct and fitting. In this example, the layer key *mhvac* is assigned to the system definition. In the Mech-Elec layer key style, this points to the layer M-Hvac. With this setup, ducts and fittings will both be placed on the layer M-Hvac. An override SUPP is then added to the third part of the layer name or the Minor field. As seen in the second example, an override replaces anything in that field. In this case, there is nothing assigned to the Minor field in the layer name M-Hvac, so the override is simply appended to the layer name. Ducts and fittings will be placed on the layer M-Hvac-SUPP. How your office sets up their system definitions will determine whether ducts and fittings end up on the same layer.

The system definition also controls the connectivity between parts. This is a bit simplistic. In truth, the connector style controls whether parts will connect or not. However, you probably will never need to modify the HVAC Connector style.

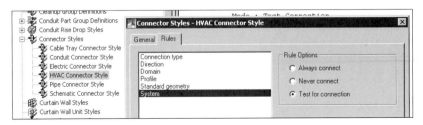

By default, when you add a duct to another duct, the connector style tests to see if the systems are the same. If they are, the connection is valid and the ducts are allowed to connect. If they are not, settings on the Building Systems Layout Rules tab of the Options dialog box take over and you are given the choice of changing the objects to the same system or placing them in the drawing without a valid connection. There are times when you will want a valid connection to occur between systems. For example, you may set up a building model that has a 1$^{st}$ floor air-supply system definition and a 2$^{nd}$ floor air-supply system definition. If you draw ducts with these systems in the same drawing, you will want them to connect. This is the function of the System Group. Systems assigned the same system group will connect.

This lesson contains four exercises. In the first exercise, you will create a system definition to orient you to the variables stored in the definition. In the second exercise, you apply the system definition to an existing duct and change the system of the entire branch. The third exercise works with the system assignment on the individual connectors of a VAV unit. The fourth exercise demonstrates how to apply global layer key overrides that affect all of the automatic layering functions.

Each exercise has a drawing file associated with it.

This lesson contains the following exercises:

- Creating System Definitions
- Changing the System for a Duct Branch
- Changing the System for MvParts
- Using Layer Key Overrides

When you complete the exercises in this lesson, you will have created an HVAC system definition. You will also have applied the system definition to a duct branch. At the end of each exercise, there is a screen shot of what your finished drawing should look like.

## EXERCISE 1: CREATING SYSTEM DEFINITIONS

System definitions apply common variables to a set of Building Systems objects. Building Systems ships with a standard set of system definitions for HVAC and Pipe. For example, the HVAC system definitions include return, supply, exhaust, and more. The Pipe System Definitions include chilled water, condenser water, hot water, and more. These system definitions are loaded when you start a drawing with a Building Mechanical template, such as the *Aecb Building Model (Imperial) 3.dwt*. If you cannot find a system definition that meets your needs, you can create one. Although you can create a new definition from scratch, it is always easier to copy an existing definition that is close to what you need and modify its values. You can create system definitions at any time, but it is easier to maintain your drawing if you create the system definitions first.

You have the ability to work with multiple systems within the same drawing. You can also group similar systems into system groups. System definitions are style-based, and therefore are accessed through the Style Manager.

In this exercise, you will be modifying a sample HVAC system definition that was loaded with the AECB Building Model template in order to create a new system definition.

 **Open:** *04 System Definitions.dwg*

### Create a New System Definition Based on an Existing One

1. Copy a system definition:

   a. From the Mechanical menu, click Mechanical Systems ➤ HVAC System Definition.

   b. In the left panel of Style Manager, under Duct System Definitions, click Supply.

   c. Right-click and click Copy, then right-click and click Paste.

   Select a pre-defined system that best matches the description of the system that you would like to create. This makes creating a new system definition faster, as some of your default settings are already defined for you.

   d. Right-click on Supply (2) and click Edit.

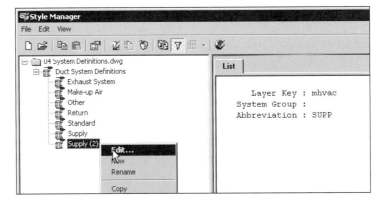

2. Specify the system definition general properties:

   a. Click the General tab and change the Name to **Plenum**.

   b. Enter **Air Handling Unit Plenum** as the description.

3. Specify the system definition design rule properties:

   a. Click the Design Rules tab

   b. Change the abbreviation to **PLNM**.

c.   Enter **Supply Air** as the System Group.

d.   Leave the layer key as mhvac.

e.   Enter **Plnm** as the Minor override.

With these settings, both ducts and fittings will be placed on the layer M-Hvac-Plnm.

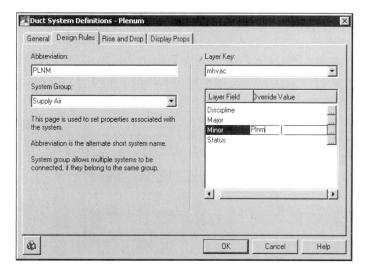

4.   Click the Rise and Drop tab and verify that the Rise and Drop Style is Supply.

5.   Adjust the display properties:

a.   Click the Display Props tab, and under Property Source select Duct System Definition.

b.   Click Attach Override, and then click Edit Display Props.

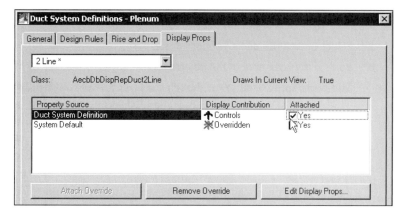

c. To select all the components at once, click contour, hold the SHIFT key, and click Rise Drop.

d. Click on any of the ByBlock entries under the color column and change the color to red.

e. Click OK four times to get back into your drawing.

**Add Ducts Between the Parts of the Air Handler to Act as a Plenum**

6. Add a duct:

a. From the Mechanical menu, click Duct ➤ Add.

b. In the Add Ducts dialog box, set the ducts system to Plenum.

c. Click the duct connector on the upper edge of the air filter as the start point of the duct.

d. Click the lower edge of the coil as the endpoint of the duct.

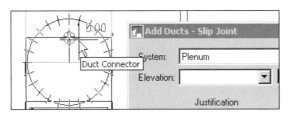

e. Press ENTER to accept the layout.

7. Repeat to add a plenum between the coil and the fan.

8. Click Yes if you are prompted to add custom parts.

9. Click the Close button to return to the drawing.

**Verify the System Definition Layering and Display Control**

10. View the layer of the plenum ducts:

a. Select one of the ducts you just drew.

b. Look in the layer area of the objects toolbar to verify the layer is M-Hvac-Plnm.

The ducts appear red (even though the layer is magenta) because you made all of the subcomponents in the system definition red.

11. View the system definitions display control of ducts:

a. Click the Plan–2 Line layout tab.

b. View the Ducts.

The ducts appear magenta because the override you specified was for the 2-Line representation of the system definition. The templates use the 2–Line representation for the RCP layouts, and the Haloed Line representation for the Plan layouts. The display representation override you set in the system definition will only be apparent in the layouts that use the 2-Line representation.

This is what your finished drawing should look like.

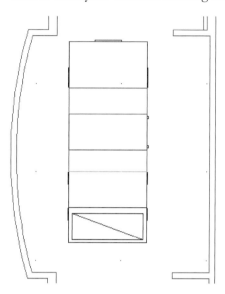

You have completed the task of creating a system definition. System definitions change the layering behavior of the ducts and fittings. Your office may want to create a drawing with office standard system definitions. Keep this drawing in a network location for everyone in the office to draw from. In addition to the sample system definitions supplied in the templates, Building Systems provides a couple of drawings that contain many more definitions. The drawings *Duct System Definitions 3.dwg* and *Pipe System Definitions 3.dwg* are installed in the …\*Program Files\Autodesk Building Systems 3\Content\Imperial\Style*s folder. Use the Style Manager to copy and paste these definitions into your drawing.

## EXERCISE 2: CHANGING THE SYSTEM FOR A DUCT BRANCH

In this exercise you will change the system of one duct to see how it travels through the entire branch. The branch in the drawing was drafted using the standard system style. The red circles in this drawing are disconnect markers. Disconnect markers show where there are breaks in the system. In this drawing you see a disconnect marker on the MvPart connections, where plumbing objects have not been attached. You also see disconnect markers in the upper right-hand corner, where there are two pieces of duct that are not connected.

**Open:** *04 Duct Systems.dwg*

### Change the System of a Branch

1. Review the duct layer:

   a. Select the main duct and note in the layer toolbar that the layer for this duct is on M-STND-Duct.

   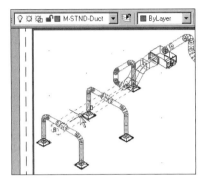

   b. With the duct selected, from the Mechanical menu, click Duct ➤ Modify.

   c. Change the System to Supply and then click OK.

2. Connecting dissimilar systems:

   a. Click Yes when the warning message appears asking you to allow dissimilar systems to connect.

      All of the connected objects are updated to reflect the supply system, even the ducts that are located on the other side of the VAV box. Where the disconnect markers appear, the ducts are not connected, so the system change stops.

   b. Select the main duct, and notice on the layer toolbar that the layer is now M-Duct_Supp, which is the specified layer applied to the supply system definition.

This is what your finished drawing should look like.

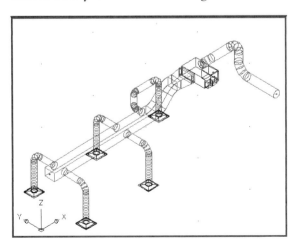

In this exercise you changed the system definition of a duct to see how it would affect the entire duct run. The system change affected all the ducts in the drawing that were connected. It is important to note that the system change extended through the VAV MvPart. The ducts in the upper right corner of the drawing were not changed because of the break in continuity between the ducts. You can use this fact to your advantage. If you do not want a system change to travel through a particular MvPart, move the MvPart an inch or so to break the continuity before you change the system definition of the duct.

When you alter system definitions for a duct, the system changes the connector on the MvPart connected to the modified duct. This is how the change is translated through the part. The next exercise takes you through the same process, but you modify the MvPart instead of the duct.

### EXERCISE 3: CHANGING THE SYSTEM FOR MVPARTS

Unlike ducts and automatic fittings, a system definition is not assigned to an entire MvPart, but rather is assigned to each connector on the MvPart. This makes sense, because although you would never see a duct with a return system at one end and a supply system at the other, you might like to see a fan MvPart with a supply connector on one side and a return connector at the other.

When you add an MvPart to the drawing it is created with no system assignments on the connectors by default. This allows you to connect a duct with any system to any of the connectors on the MvPart. Once you attach a duct with a system to an MvPart, that connector assumes the system of the duct attached to it. You can also manually set the connectors on an MvPart through the MvPart Properties dialog box.

In this exercise, you control the system definition on the connectors of MvParts. You do this manually by changing the properties of the MvPart, and you do this automatically by adding a duct to the MvPart.

 **Open:** *04 MvPart Systems.dwg*

### Change a Coil Connector and Add a Duct

1. Access system properties for an MvPart:

   a. From the MEP Common menu, click MvParts ➤ MvPart Properties.

   b. Select the coil on the lower-right VAV box and press ENTER.

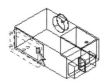

c.  Click the Systems tab.

Here you see all of the connectors that are built into the coil. For this coil, there are two connectors that belong to pipes, and two connectors that belong to ducts. By default, all of the connector systems are set to "none" when an MvPart is inserted into the drawing. Thus, you can specify any duct or pipe system to the connectors that you want.

2.  Define systems for the connectors:

a.  Click under System for Connector 1 and you will get a drop-down list. Select Return for the system assigned to the connector.

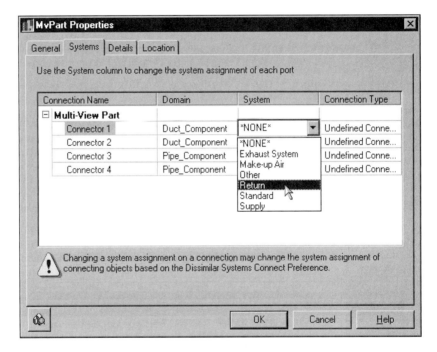

b.  Repeat to assign a Return system to Connector 2.

c.  Click OK.

3.  Draw a duct from the Coil:

a.  From the Mechanical menu, click Ducts ➤ Add.

b.  Change the System to Supply.

c. Pick the open connector on the coil as the start point for the duct. Drag your cursor to the left and pick a second point to end the duct. Press ENTER to end the command.

Just as the connector passes the size information into the Add Duct dialog box, the system information is passed on as well. The duct system changes to return once you select the coil connector with the return system assigned to it.

### Add a Duct to the VAV Box to Assign a System to the Connector

4. Review the VAV box connectors:

a. From the MEP Common menu, click MvParts ➤ MvPart Properties.

b. Select the VAV box on the lower right and press ENTER.

c. Click the Systems tab.

All the connectors are set to None.

5. Add a duct to the intake of the VAV:

a. From the Mechanical menu, click Ducts ➤ Add.

b. Change the System to Supply.

c. Select the round connector on the upper-right face of the VAV box as the start point.

d. Pick a second point for the duct and press ENTER.

6. Review the VAV box connectors:

a. From the MEP Common menu, click MvParts ➤ MvPart Properties.

b. Select the VAV box and press ENTER.

c. Click the Systems tab.

Connector 3, that you just added the Supply duct to, has inherited the Supply definition from the duct.

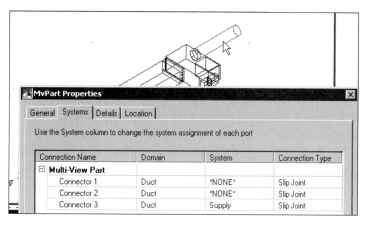

## Change the Duct System and Review the Connectors

7.  Change the duct system:

    When you change the system of a duct, the system of any connected MvParts also changes.

    a.  Select the rectangular duct segment you just drew, right-click, and click Duct Modify.

    b.  Change the System to Supply.

    c.  Click OK.

        If a warning message appears asking you to allow dissimilar systems to connect, click Yes.

8.  Review your changes:

    a.  Select the coil again, right-click, and click MvPart Properties.

    b.  Click the Connectors tab.

        Notice that Connectors 1 and 2 are now assigned to the supply system.

    c.  Click OK.

This is what your finished drawing should look like.

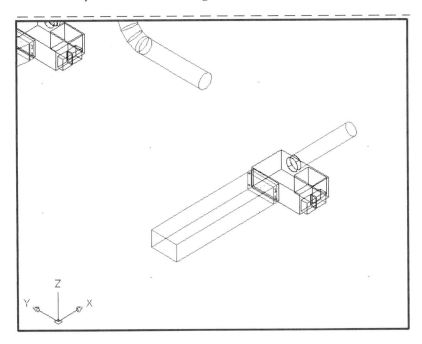

In this exercise, you changed the system definition of an MvPart's built-in connectors, drew duct segments, and modified the system for all of the connected pieces.

The ability to change a duct system is a powerful tool. However, as you have seen here, be careful that you do not accidentally re-assign a system after the design is created. If you have to change a system after the design is drawn, you can control how far the system re-assignment travels by moving a piece of the system (or "breaking" the system) at the place where you want the system change to stop.

The other situation where you will have to manually control the system assigned to a connector is if you add fittings manually to the drawing. The Add Fitting dialog box lacks the system assignment control, so you will have to do this manually for fittings you add to the drawing. The alternative is to place the fitting in the branch, then use the duct modify to change the branch to a different system, and then change it back again. If you do this, the manually added fitting will then layer key to the final system as well.

### EXERCISE 4: USING LAYER KEY OVERRIDES

The default layer naming convention within the Building Systems software is chosen when you install the software. There are several different layer standards and layer key styles provided with the software such as the European BS1192. The exercises in this book all have been set up with the Mech-Elec layer key style which uses the Mech-Elec layer standard. These components are based on the AIA layer standard that has a specific format to the layer name. Each layer name consists of four fields, a Discipline designation that is one character, a Major field, a Minor field, and a Status field. The last three fields hold four characters each.

In the first exercise in this lesson, you created a system definition. You assigned an override to the layer name within the definition. This allowed different systems to layer the ducts and fittings differently. Global layer key overrides will work in the same way, but will affect everything you place in the drawing. In this lesson you will set a global override and place equipment and ducts in the drawing. You will also use the Remap Objects Layer tool to change the layers of the objects.

This exercise assumes you have worked through the other exercises in this chapter and are familiar with adding a duct and MvParts to the drawing.

 **Open:** *04 Layer Key Overrides.dwg*

### Set a Global Override

1. Set an "-Exst" Override to the status field of the layer name:
   a. From the Desktop menu, click Layer Management ➤ Layer Key Overrides.
   b. Click the ellipses button to the right of the Status field.

c. From the list of Pre-Specified values, Click Exst.

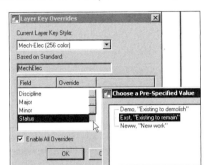

d. Click OK twice to return to the drawing.

## Add a Coil and Ducts to the Drawing with the Override

2. Add a Coil:

a. From the Mechanical menu, click MvParts ➤ Add.

b. Add a 28 x 16 in. VAV Reheat Water Coil to the connector 2 of the VAV unit in the lower part of the drawing.

The VAV coils are in the MvParts Catalog under VAV units ➤ VAV Reheat Coils. Use the command line basepoint toggle to match the connectors.

3. Draw ducts:

a. From the Mechanical menu, click Ducts ➤ Add.

b. Add a supply duct from the coil extending to the left.

c. Connect the air terminal to the duct you just drew.

Remember to check that the duct connection preference is set as takeoff, not tee, before you connect the air terminals.

## Review the Layers Assigned to the Parts

4. Review the layer for the coil:

a. Select the coil you placed in the drawing.

b. Look in the layer area of the objects toolbar to see that the layer is M-STND-Eqpm-Exst.

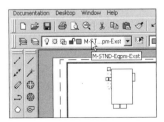

5.  Review the layer for the ducts:

    a.  Select the duct you placed in the drawing.

    b.  Look in the layer area of the objects toolbar to see that the layer for the duct is M-Duct-Supp-Exst.

        The override applies to all objects you add to your drawing. The layers M-STND-Eqpm-Exst and M-Duct-Supp-Exst were created on the fly. The layers were created using settings stored in the layer key for the object added.

        To follow our duct example; when you added a duct with the –Exst override, the duct looked in the existing layer list for a layer called M-Duct-Supp-Exst. Not finding one, it used the settings from the layer key style to create one. There isn't a separate layer key for each override, so M-Duct-Supp and M-Duct-Supp-Exst will be assigned the same color and linetype. You can manually change the properties of layer M-Duct-Supp-Exst with the Layer Manager to gray and dashed. As with all layer keying, once the layer exists in the drawing, the layer key will not change it. The layer key and layer key style only serve to create new layers if they do not exist in the drawing.

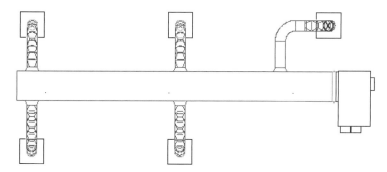

## Remap Object Layers to Their Original Layer Settings

6.  Reset the global override:

    a.  From the Desktop menu, click Layer Management ➤ Layer Key Overrides.

    b.  Select and Delete "-Exst" from the Status field.

    c.  Click OK twice to return to the drawing.

7.  Remap the ducts, fittings and flex duct:

    a.  From the Desktop menu, click Layer Management ➤ Remap Object Layers.

    b.  Select the ducts, fittings, and flex duct in the drawing and then press ENTER.

    c.  Type **O** (for Object) at the command line to remap back to the original objects layer keys.

        The ducts, fittings, and flex duct remap back to their original layer keys and disregard the layer settings of the supply system. Ducts are now on layer M-STND-Duct, fittings on M-STND-Dfit, and flex duct on M-STND-Dflx.

8.   Remap the air terminals:

   a.   From the Desktop menu, click Layer Management ➤ Remap Object Layers.

   b.   Select the air diffusers and then press ENTER.

   c.   Type **O** at the command line.

   The air terminals remap back to M-EQPM. All MvParts will remap back to M-EQPM. MvParts in the drawing do not retain a hard-coded layer key. The layer key is defined in the parts catalog and used only when the part is added to the drawing. To remap MvParts, use Quick Select to select the parts by type, then use the layer key option of the Remap tool, entering at the command line the layer key you want to remap that set of MvParts to.

This is what your finished drawing should look like.

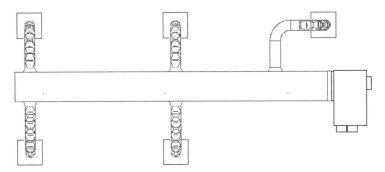

In this exercise, you applied a global layer override and remapped objects. In practice, you can use the layer key override and the remap layer tool together. When you remap object layers, any overrides you have in place are used by the remap tool. For example, if you had an existing mechanical model created from Building Systems objects and needed to change all the objects to a layer with the extension – Exst, you could set the global layer key override, and then remap the objects, changing them to layers matching their layer key, but with the –Exst extension.

In Lesson 5, "Working with System Definitions and Layer Keys," you completed four exercises: Creating System Definitions, Changing the System for a Duct Branch, Changing the System for MvParts, and Using Layer Key Overrides. You have used both the controls and overrides stored by the system definition. You have also applied a global layer key override.

# KEY CONCEPTS: CREATING A MECHANICAL PLAN

Many background settings are stored on the tabs that Building Systems adds to the AutoCAD Options dialog box.

The Duct Layout Preferences control the default type of duct you add as well as the default fittings that will be automatically placed in the drawing by adding ducts.

MvParts are stored in a catalog. The paths to the catalog are stored in the Building Systems tabs of the Options dialog box.

MvParts have connectors that store the size and system definition. The size and system are passed on to a duct when it is added to the connector on the MvPart.

MvParts are made up of AutoCAD blocks.

As you add MvParts to the drawing, there are command line toggles that help you place the parts in the drawing.

As you add ducts, fittings are placed automatically.

You can add fittings manually that are not part of the automatic layout of the ducts and fittings.

The Compass is a tool to help you route ducts through the building model.

System definitions are responsible for layering ducts and fittings.

System definitions are used by the software to test if two ducts should connect or not.

Each Building Systems object is assigned a layer key. The layer key is mapped to a layer key style to determine the automatic layering of objects.

A layer key override modifies one of the four layer name fields for all objects placed in the drawing.

# Reviewing Your Designs

After you read this chapter, you should understand:

- how to create a schedule for the VAV boxes in a drawing
- how to add schedule data to the schedule that you created
- how to add and delete columns in your schedule
- how to edit the schedule data
- how to edit the schedule table style to reformat the schedule table

## REVIEWING YOUR DESIGNS

Towards the end of your project it is important that you review your designs. Autodesk Building Systems enables you to perform a thorough review before handing your finished drawing off to a client or customer. To perform a review you will check for breaks in your duct or pipe runs, check for structural interferences, and create a schedule of the parts in your drawing.

This chapter covers the following features of Building Systems and Architectural Desktop, which enable you to review your designs:

- Disconnect Markers and Interference Detection
- 2D Sections
- Schedules

## USING DISCONNECT MARKERS AND INTERFERENCE DETECTION

Disconnect Markers and Interference Detection are both features of Building Systems. Disconnect Markers alert you to connectors or connections that are not valid because the nodes are misaligned, or the systems of the two pieces do not match. Disconnect Markers are either on or off—the value of this toggle is stored in the drawing. Disconnect Markers appear as red circles in the drawing. The red circles can appear under several conditions. One condition is when there is a connector on a Building Systems object that is 'open', meaning that a connection exists that has nothing connected to it. A Disconnect Marker also appears at broken connections, for example, when two ducts are close to one another, but not touching at the centerline.

Another condition that causes a Disconnect Marker to appear is when the properties contained in the duct connector styles do not match. If two ducts contain different styles, shapes, or sizes, the connection between the two ducts is not valid.

You can see the specified duct connector style in the Style Manager. From the Desktop menu, click on Style Manager and highlight Connector Styles. Select the HVAC Connector Style to see a list of the connector style settings.

The conditions by which connections are analyzed are stored in the HVAC Connector style.

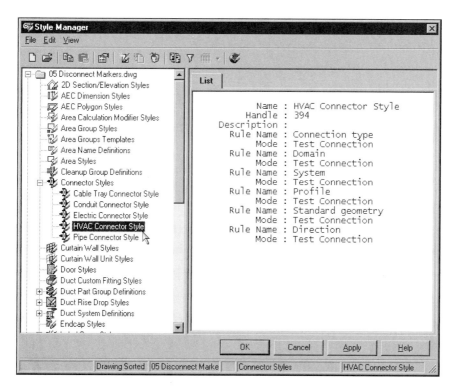

Interference Detection appears as a red warning when a Building Systems object such as a duct or pipe fitting crosses through another Building Systems object or structural member. Structural members are Architectural Desktop objects such as beams and columns.

Interference Detection does work through externally referenced files (xrefs). For example, beams in an xref interfere with ducts in the active drawing.

## UNDERSTANDING TWO-DIMENSIONAL (2D) SECTIONS

Sections are Architectural Desktop features that have been adopted by Building Systems. You use 2D section lines to create 2D section objects. You create 2D section objects to view where the different Building Systems objects are in relation to one another. Once the 2D section object is created, you can use different drawing views, such as isometric and plan, to review your drawing. This is an important step in identifying any places where objects may obstruct one another. Examples of when a 2D section may be useful would be for mechanical rooms, HVAC layouts, and roof or wall penetrations.

When using this feature be sure that all of the layers are turned on for the objects that you would like to view. You can also use Building Systems's two-line edge toggle command to hide unwanted edges when viewing two-line displays. If you want to add depth to your 2D section object you can create subdivisions within your section and apply different lineweights to each subdivision.

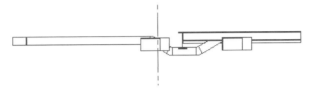

A 2D section object is linked to your drawing, and therefore can be updated when drawing changes are made.

## UNDERSTANDING SCHEDULES

Schedules are a feature of Architectural Desktop that enable you to produce a detailed list of all of the objects in your drawing. The information that you decide to include in your schedule can be placed into a schedule table in the drawing. Any changes that you make in your drawing are automatically reflected in your schedule table. Schedules may also be exported into a Microsoft® Excel speadsheet.

There are three different style definitions needed to create a schedule table, as well as schedule tags, which use the same information that a schedule uses.

The three styles and definitions that you will work with in this chapter are:

- Schedule Data Formats
- Property Set Definitions
- Schedule Table Styles

Schedule Data Formats store the variables of how the text or numbers are formatted in the schedule table cells that are specified for each property you create.

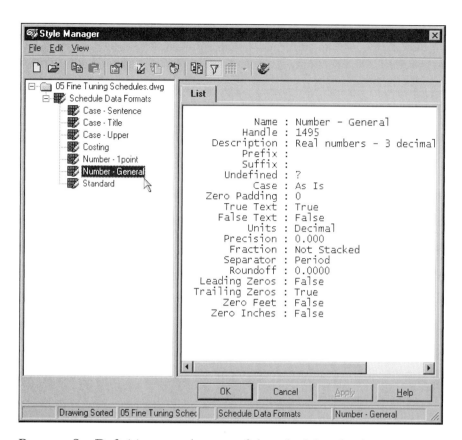

Property Set Definitions are the guts of the schedules. As their name implies, they are a group of properties that define an object. Properties come in two types: Automatic and Manual. Automatic properties read values such as style name, or dimensions from the AEC objects. Once placed in a schedule you do not have direct control of a property through the schedule, but must change the referenced object to change the value in the table. Manual properties are filled in by you, and may be edited directly in the schedule table. Each column in a schedule is an individual property stored in a Property Set Definition. Property Set Definitions may be assigned to styles or to the individual objects. Property Set Definitions are the holder for the information that will be used in the schedule. This information is referred to in the software as the Schedule Data, so any time you see Schedule Data, think Property Set Definition, and vice versa.

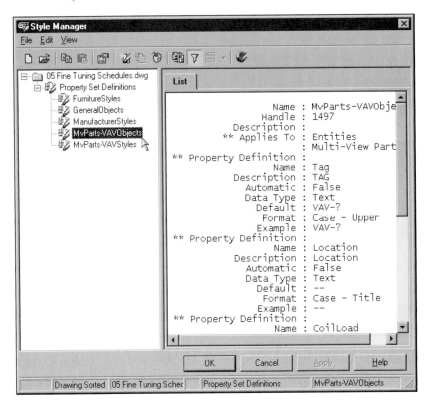

Schedule Table Styles hold the organization of the schedule table. The Schedule Table Style determines the order of the columns, the order of the rows, which property is assigned to each column, as well as any overrides that you want to assign to change the style of the text in the headers.

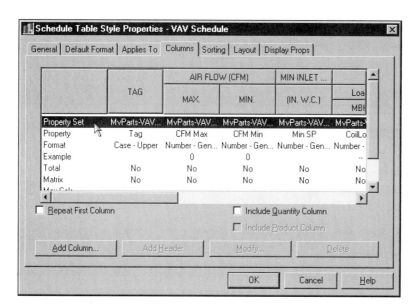

These are the three major components needed in creating a schedule table. There are also Schedule Tags, but Schedule Tags are not necessary in the drawing to produce a schedule. Schedule Tags are a special kind of Multi-view Block (not to be confused with an MvPart). A Schedule Tag is created with an attribute that looks at the schedule data attached to the object and reads the properties out of the schedule data, and then displays that data as its own attributes. When you place a Schedule Tag on an object, the program also assigns the schedule data to the object, but you can do this manually as well.

## LESSON OBJECTIVES

This chapter contains three lessons. The lessons are broken down into one or more exercises. Most exercises have a corresponding drawing file that is located on the CD-ROM included with this book.

Lessons in this chapter include:

- **Lesson 1: Using Building Systems to Review Your Design**

  In this lesson you use the two primary Building Systems tools (Disconnect Markers and Interference Detection) to review your design.

- **Lesson 2: Checking Interference Using Sections**

  In this lesson you add a Section Line, adjust the Section Lines Properties, and then generate a 2D Section Object.

- **Lesson 3: Scheduling MvParts**

  In this lesson you create a schedule, attach schedule data, edit the schedule, and edit the schedule table style.

## LESSON 1: USING BUILDING SYSTEMS TO REVIEW YOUR DESIGN

This lesson contains two exercises, Using Disconnect Markers and Showing Interference with Beams. Each exercise has a drawing file associated with it. When you complete the exercises in this lesson you will have used the two primary Building Systems tools, Disconnect Markers and Interference Detection, to review your design.

To accomplish this task you will be working with two drawings: one that contains invalid connections, and one that contains structural interference. At the end of each exercise there is a screen shot of what your finished drawing should look like.

This lesson contains the following exercises:

- Using Disconnect Markers
- Showing Interference with Beams

### EXERCISE 1: USING DISCONNECT MARKERS

A Disconnect Marker can appear when there is a connector on a Building Systems object that is 'open', meaning that a connection exists that has nothing connected to it. You see an example of this in the exercise drawing where the coil connectors are not connected to the pipe, and when the return grill diffuser connections are open to the ceiling plenum because there is no duct attached. In this exercise you turn on Disconnect Markers to evaluate where the duct system is not properly connected.

 **Open:** 05 Disconnect Markers.dwg

### View the Broken Connections

1.  Show the Disconnect Markers:

    a.  From the MEP Common menu, click Utilities ➤ Disconnect Markers.

    b.  At the command line, enter **Y** to show the disconnect markers and then press ENTER.

    c.  At the command line enter **OBJRELUPDATE** and press ENTER twice.

        At the lower right, a Disconnect Marker appears at the connection between the round 16-inch duct and the round 14-inch flex duct because their diameter does not match.

### Fix the Broken Connections

2.  Fix the disconnect by changing the duct diameter:

a. Select the duct at the lower right, right-click, and click Duct Modify.

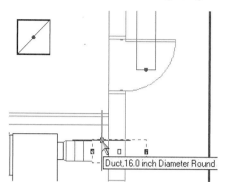

Duct,16.0 inch Diameter Round

b. In the Modify Duct dialog box, change the Diameter to 14" and then click OK.

c. At the command line, enter **REA** to regenerate all.

Notice that the Disconnect Marker goes away because you have fixed the invalid connection caused by two dissimilar duct sizes.

At the top of the drawing you should see a Disconnect Marker where the green flex duct meets the fitting to the VAV box. This disconnect is caused because the systems of the two connectors do not match. The system information is also stored at the connector, so two parts can be the same size, but will still not connect if they have different systems assigned to them.

3. Fix the disconnect by changing the duct system:

a. Select any of the green ducts, right-click, and click Duct Modify.

b. In the Modify Duct Dialog box, change the System from Standard to Supply and then click OK.

c. In the 'Allow Dissimilar Systems to Connect' warning dialog box, click Yes.

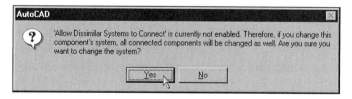

d. At the command line, enter **REA** to regenerate all.

Notice that the Disconnect Marker goes away because you have fixed the invalid connection.

This is what your finished drawing should look like.

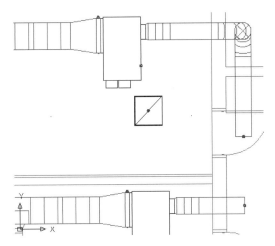

Disconnect Markers are stored in the drawing. If you save the drawing with the Disconnect Markers turned on, then the next person that opens the drawing will see them. We don't recommend that you work with Disconnect Markers turned on because of its slower drawing performance.

In this exercise you turned on Disconnect Markers to view the invalid connections. You also fixed some broken connections by changing the size and system of ducts, and matching the size and system of the connected duct.

### EXERCISE 2: SHOWING INTERFERENCE WITH BEAMS

In this exercise you will turn on Interference Detection to evaluate any conflict between ducts and structural members in the drawing.

 **Open:** *05 ABM Interference.dwg*

#### View Interferences in Your Drawing

1.  Access the Interference Detection command:

    a.  Click in the left viewport to make it active.

    b.  From the Tools menu, click Options.

    c.  Click the Building Systems Layout Rules tab.

2.  Run Interference Detection:

    a.  Select the Alert check box under Building Systems Interference Detection.

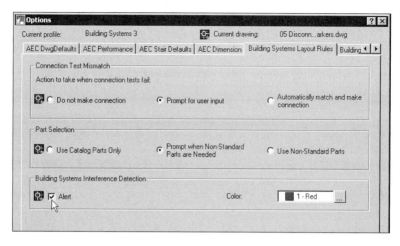

b.  Click OK.

c.  At the command line, enter **REA** for regenerate all and then wait.

    It is recommended that you do not work with Interference Detection turned on due to its slow drawing performance.

This is what your finished drawing should look like.

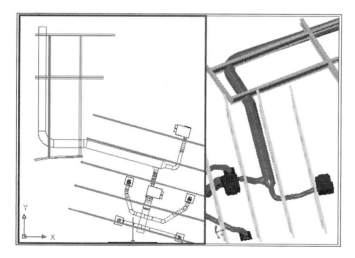

In this exercise you turned on Interference Detection to view the structural interferences. The plan view shows the four areas where the ducts and beams are in conflict–two flex ducts to the web joist and two places where the main duct crosses through an I-beam. You can move the main duct trunk or resize it to be an oval or shallow rectangle to avoid the interference.

The next lesson shows you how to create a temporary section object to check for interference.

## LESSON 2: CHECKING INTERFERENCE USING SECTIONS

On a simple model, looking for interference may be as simple as changing the view to a side view and looking for conflicts. Turning on interference checking is also an option, however if you just need to check a single point, you can do that using the Architectural Desktop 2D Section tool. The following exercises take you through creating a 2D section.

To accomplish this task you will be adding a Section Line, adjusting the Section Line Properties, and then generating a 2D Section Object. At the end of each exercise there is a screen shot of what your finished drawing should look like.

This lesson contains the following exercises:

- Adding and Adjusting a Section Line
- Generating a 2D Section
- Working with a 2D Section

### EXERCISE 1: ADDING AND ADJUSTING A SECTION LINE

In this exercise you will place a section line in the drawing and edit the section line properties. The changes that you make to the section line will be used in Exercise 2, "Generating a 2D Section," where you adjust the 2D section properties. There are two points in the exercise drawing to help you place the section line. The points appear as round circles with an X inside them.

 **Open:** *05 Placing Section Line.dwg*

**Add the Section Line**

1.  Add the section line:

    a.  From the Documentation menu, click Sections ➤ Add Section Line.

    b.  Verify that only the Node Osnap is turned on and pick the lower node and the upper node and press ENTER.

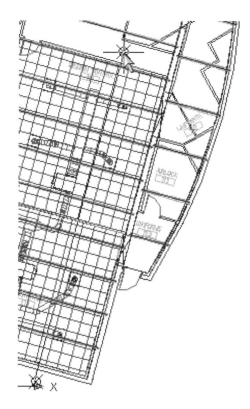

c. At the command line, enter **10'** for length and press ENTER. Press ENTER again to accept 10' for the height.

The length also represents the depth of the section line.

The section identification bubbles are multi-view blocks: to edit their attributes, select the bubble, right click to access the MvBlock Reference Properties dialog box. On the Attributes tab you can change the attributes for the MvBlock.

**Note:** You can also use the Documentation menu to select Documentation Content ➤ Section Marks to access more section marks from the DesignCenter. These section marks have the sheet reference in the bubble.

### Adjust the Section Line Properties

2. Adjust the Section Line Properties:

   a. Select the section line, right-click, and click Section Line Properties.

   b. Click the Dimensions tab and select Use Model Extents for Height.

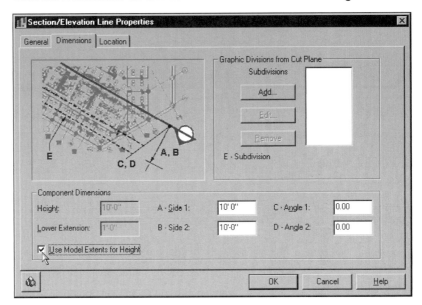

   This overrides the 10' height that you accepted when you were adding the section line. Selecting the Use Model Extents for Height check box guarantees that the section line will cut through all of the objects in the drawing, no matter what their height.

3. Add three subdivisions from the cut plane:

   The cut plane is established by the point you pick when you add the section line and defines where the section will be cut through the building. Adding cut plane subdivisions enables you to define specific distances from the cutting plane in order to graphically divide the space.

   a. Under Graphic Divisions From Cut Plane Subdivisions, click Add.

   b. In the Add Subdivisions dialog box, enter **3'** and then click OK. Click Add to define the second subdivision.

   c. Enter **5'** for the second subdivision and click OK. Add the third subdivision at **8'** and then click OK twice.

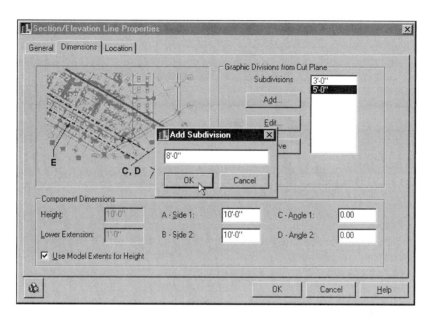

This is what your finished drawing should look like.

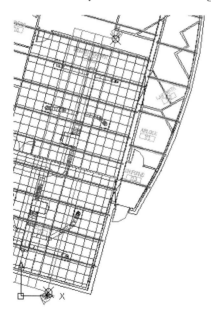

You have completed the task of creating a section line, and adding subdivisions to that section line. In the next exercise you will generate a 2D section.

## EXERCISE 2: GENERATING A 2D SECTION

In this exercise you generate a 2D section and then adjust the section properties to display different colors for the subdivisions that you created in the previous exercise.

**Open:** *05 Generating Section.dwg*

### Generate a 2D Section

1. Select the drawing objects to use in the 2D section:

   a. Zoom to the extents of the drawing and select the section line, right-click, and click Generate Section.

   Do not select the section mark bubbles at each end of the section line. You need to select the section line itself. The bubbles are not part of the section line.

   b. Under Selection Set, click Select Objects.

   c. Using a window, select the entire building and press ENTER.

   In the Generate Section/Elevation dialog box, verify that 71 items were selected.

2. Generate a 2D section:

   a. Under Placement, select New Object and then click Pick Point.

   b. Pick a point to the lower right of the building and then click OK.

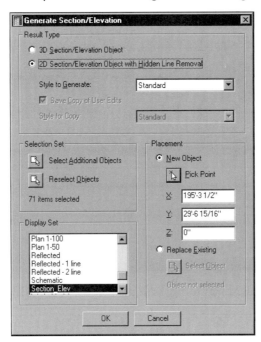

The Generating Section/Elevation dialog box shows that the program is cutting the section.

You now have a 2D section in your drawing. The next part of this exercise uses the display properties of the 2D section to change the colors of the different subdivisions that you created in Exercise 1.

### Adjust the Colors of the Subdivisions in the 2D Section

3.  Access the display properties of the 2D section:

    a.  Pan to the right in your drawing and select the 2D section object, right-click, and click Edit 2D Section/Elevation Style.

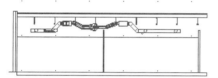

    b.  Click the Display Props tab, and under Property Source select 2D Section/Elevation Style.

    c.  Click Attach Override and then click Edit Display Props.

4.  Change the colors of the subdivisions:

    a.  On the Layer/Color/Linetype tab, select Subdivision 1.

    b.  Under Color, click By Block and change the color to **30**.

    c.  Change the Subdivision 2 color to **52** and the Subdivision 3 color to **102**, and then click OK twice.

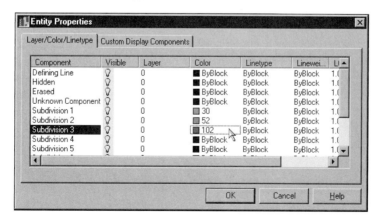

You may have to regenerate the drawing to see the new colors. You regenerate the drawing by entering **REA** at the command line.

This is what your finished drawing should look like.

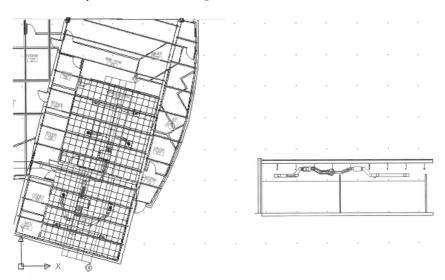

You established Cut Plane Subdivisions using the Section Line properties in Exercise 1. Now that you have created and used subdivisions, you can see how they are important in visually distinguishing different aspects of your drawing when using a 2D section. The next exercise expands the concept of using display properties by creating a component called 'Lighter,' that you can use in the 2D section and control with the display properties of the 2D section.

### EXERCISE 3: WORKING WITH A 2D SECTION

In this exercise you will adjust the style and the display properties of the 2D section object. You also work with the two commands for 2D sections, add and merge. The 2D section object has many different subcomponents that you can control by adjusting the display properties like you did in the last exercise. In this exercise you create a new component in the 2D section object style, adjust the display properties, and then add some linework to the component that you created.

**Open:** *05 Working 2D Section.dwg*

#### Create a New Component in order to Draw on the Section

1. Create a new component:

   a. Select the 2D section object, right-click, and click Edit 2D Section/ Elevation Style.

b.  On the Components tab, click Add, and enter **Lighter** for the new Component name.

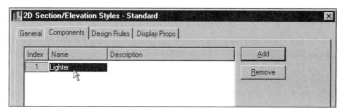

If you would like to learn more about how components work in Architectural Desktop, click the Help button.

2.  Edit the display properties of the component:

a.  Click the Display Props tab, and under Property Source select 2D Section/Elevation Style.

b.  Click Edit Display Props, and on the Layer/Color/Linetype tab, select Lighter, and change its color to **200**.

c.  Select the component called Erased, turn off the visibility, and then click OK twice.

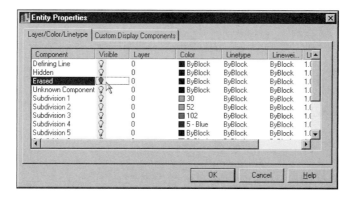

## Draw Linework to Add to the Section

3.  Draw some linework to add to the section:

a.  Turn off Osnap and use two lines to draw a cross on the left round duct that is coming toward you.

This cross provides a visual to designate this duct as a supply duct.

b.  Select the 2D section, right-click, and click Merge Linework.

c. Pick the two lines that you just drew and press ENTER.

You must pick the lines—you cannot use a window, and crossing does not work inside this command.

4. View the subcomponents contained in your drawing and place the two lines on the component that you created called Lighter:

a. Press **F2** to bring up the AutoCAD text Window.

All of the subcomponents of the 2D section object are listed here. The command line is prompting you to define which subcomponent that you want to place these lines on.

b. Enter **14** to place the two lines on the component that you created called Lighter, and press ENTER.

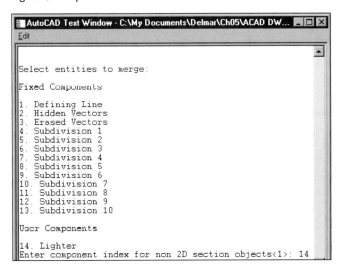

c. Press **F2** to close the AutoCAD Text Window.

The lines are merged into the 2D section.

## Change Linework to be on a Different Component

5. Change the linework's component:

a. Select the 2D section, right-click, and click Edit Linework.

b. Pick the interior lines on the round duct that is located to the right of the duct that you added a cross to, and press ENTER.

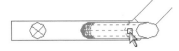

c. At the command line, enter **3** to place these lines on the 'Erased' component of the 2D section.

Since this component is turned off, these lines do not display.

This is what your finished drawing should look like.

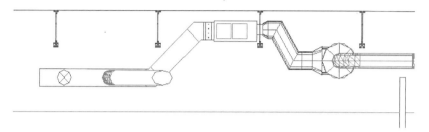

In this exercise you created a component within a 2D section style, and then added linework to the component to use as a visual drawing reference.

In Lesson 2, "Checking Interference Using Sections," you merged and edited linework in a section. It is important to note that the next time you update a section that you edited, the edits that you made appear discarded because they turn into another 2D section. Therefore, you must merge the two 2D sections together after you update them.

You did not use the Design Rules tab on the 2D Section/Elevation Styles dialog box in this lesson. The Design Rules tab enables you to select anything by the AutoCAD Color Index (ACI) in the model and then assign it to one of the display components.

 **Caution:** Do not associatively hatch a 2D elevation or section object. You can, but it will cause a heavy performance hit.

In practice, it can be more efficient to generate the section and use it as a template or guideline and then draw directly on top of the section. You can also use the Boundary Creation command (enter Bpoly at the command line) to pick points and generate polylines within the section. You can use these polylines to hatch associatively, but remember to move the section object from under the polylines first to keep from inadvertently hatching the section object.

When you plot drawings, you can change the layer of the section object to be on a layer that does not plot. Therefore, if the design changes, you still have a quick reference to grip-stretch the hatches to, and therefore you won't have to spend time creating extension lines.

## LESSON 3: SCHEDULING MVPARTS

This lesson contains five exercises: Creating a Schedule, Adding Schedule Data, Adding a Column to the Schedule, Editing Schedule Data, and Editing the Schedule Table Style. Each exercise has a drawing file associated with it. When you complete the exercises in this lesson you will have created a schedule, attached schedule data, added a column, edited the schedule, and edited the schedule table style. To accomplish these tasks you will be using the Architectural Desktop Schedule command, including schedule data format styles, property set definitions, and schedule table styles. At the end of each exercise there is a screen shot of what your finished drawing should look like.

This lesson contains the following exercises:

- Creating a Schedule
- Adding Schedule Data
- Adding a Column to the Schedule
- Editing Schedule Data
- Editing the Schedule Table Style

### EXERCISE 1: CREATING A SCHEDULE

In this exercise you will import a predefined schedule table style into your drawing using Style Manager. Schedule table styles define how the formatting of your finished schedule will appear. In Architectural Desktop, schedule table styles are built into the templates. However, there are no schedule table styles built into the Building Systems templates.

You also use the Quick Select command to select all of the VAV boxes in your drawing to use in your schedule. You then place the schedule into your drawing.

 **Open:** *05 Placing Schedules.dwg*

#### Copy a Schedule Table Style into Your Drawing

1. Review the style information in the drawing:

   a. From the Desktop menu, click Style Manager.

   b. In the left panel of Style Manager, select and expand the *05 Placing Schedules.dwg* folder.

   c. Scroll down the list of styles and definitions in the drawing. Notice that there are no schedule table styles, property set definitions, or schedule data formats.

   If there were any schedule table styles in the drawing, a box would appear with a plus (+) sign next to the style.

   d. Click OK to return to the drawing.

2. Import a schedule table style from another drawing:

   a. From the Mechanical menu, select Tags and Schedules ➤ HVAC Schedule Tables.

   This command opens the Style Manager and opens the drawing *ABM Duct Schedule Tables (Imperial) 3.dwg* in the Style Manager that contains sample mechanical schedules for you to use. This drawing is not open in your AutoCAD session, but only open within the Style Manager for you to copy a style from.

3. Select the schedule table styles:

   a. In the left panel of Style Manager, select Schedule Table Styles and then click the Filter Style Type icon.

   The Filter icon looks like a funnel. After clicking the Filter icon you will see that only the schedule table styles are shown in the two drawings.

   b. Expand the schedule table styles in the *ABM Duct Schedule Tables (Imperial) 3.dwg* folder.

   c. In the left panel, select Schedule Table Styles to view a list of the available styles in the right panel of Style Manager.

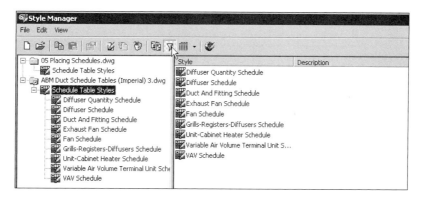

4. Drag a schedule table style into the *05 Placing Schedules* drawing:

   a. In the right panel, select VAV Schedule and drag it to the left, placing it into the *05 Placing Schedules* drawing.

   You place the VAV Schedule into the *05 Placing Schedule* drawing by highlighting the VAV Schedule, and while it is still highlighted, hovering your cursor over the *05 Placing Schedules* drawing.

   b. In the left panel, expand Schedule Table Styles. You should now see the VAV Schedule listed in the drawing.

   c. Select the *ABM Duct Schedule Tables (Imperial) 3.dwg*, right-click, and click Close.

5. View the schedule table style in your drawing:

    a. In the left panel, select and expand Schedule Table Styles.

       You may need to click the Filter icon again to redisplay all of the drawing information in the left panel of the Style Manager.

    b. Expand Property Set Definitions and Schedule Data Formats.

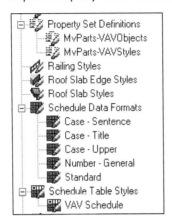

       Notice that the VAV schedule table style contains all of the necessary property set definitions—for example, MvParts-VAVObjects and MvParts-VAVStyles—and schedule data formats that it needs to create itself.

    c. Click OK to get back into the drawing.

**Note:** If you bring in a style that is already in the drawing, you are told that the style exists, and are prompted as to whether you want to overwrite the existing style. If you decide to overwrite the style, it is important to note that the subdependent styles are not overwritten as well. For example, in the above steps, you have a property set definition called MvParts-VAV Objects. The MvParts-VAV Objects property set definition would remain, and the schedule would use this instead of the property set definition that was created for the style that you overwrote. This may cause your schedule to appear incorrectly when placed in the drawing.

## Place a Schedule Table in the Drawing

When you place a schedule table in the drawing, you are prompted to select the parts that you want to place in the schedule. Because many of the things you will be scheduling are MvParts, you need to be able to select a specific MvPart type; for example, just the VAV boxes, just the coils, or just the air terminals. You can do this using Architectural Desktop's Quick Select command and then establishing the selection set before you add the schedule table. If all of the parts that go into the schedule are on the same layer, you can use the layer filter as you add them to the schedule. In this section of the exercise, you do both.

If you are working on a mid- or large-size project you may want to keep your schedules in a different drawing with the model plan drawings externally referenced into it. Architectural Desktop schedules can work through an external reference file (xref), but in order to use the layer filter appropriately you will have to append the drawing name an "|" to the layer name in the filter for the filter to work. Make sure when you first add your schedule that you select the option to scan xrefs in the Add Schedule Table dialog box.

6. Review the type of MvPart:

   a. Select one of the VAV Boxes in the drawing, right-click, and click MvPart Properties.

   b. Click the Details tab to review what type of MvPart you have selected. On a piece of paper, write down the MvPart type exactly as it appears on this dialog box—in this case it is VAV_Box.

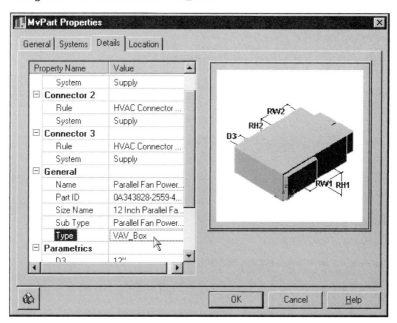

   c. Click Cancel to get back into the drawing.

7. Access Quick Select:

   a. Right-click anywhere in the drawing and click Quick Select.

   b. In the Quick Select dialog box, use Apply to select Entire drawing.

   c. For Object Type select Multi-View Part, and for Properties scroll down and select Type.

8. Select all of the VAV boxes in the drawing:

    a. In the Quick Select dialog box, verify that Equals is selected for Operator.

    b. For Value, enter **VAV_Box**.

 **Caution:** You must enter the 'type' name of the MvPart exactly as it appears on the Details tab of the Properties dialog box, or Quick Select will not recognize the MvPart in your drawing.

    c. Under How to Apply, click Include in new selection set and then click OK.

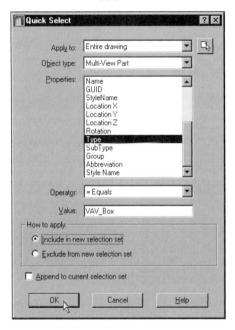

All four of the VAV boxes should be highlighted in your drawing.

9. Add the schedule table style to your drawing:

    a. From the Documentation menu, click Schedule Tables ➤ Add Schedule Table.

    b. Select the VAV Schedule.

    The VAV Schedule is the only choice because it is the only schedule table style that exists in your drawing. This is the same schedule table style that you imported at the beginning of this exercise.

    c. For Layer Wildcard, enter **M-STND-Eqpm-VVA**. Now only the objects on this layer will show in the schedule.

10. Add the schedule to your drawing:

    a. Select Add New Object Automatically and Automatic Update, and then click OK.

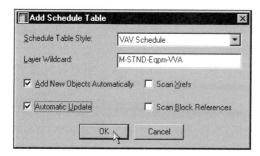

b.  You are prompted to select objects at the command line; enter **P** (for previous) and press ENTER twice to end the selection set.

Entering P at the command line specifies the program to use the set of objects highlighted from the last Quick Select command.

c.  Pick a point in the upper left-hand corner of the viewport and then press ENTER to automatically size the schedule.

The scale specified in the Drawing Setup dialog box determines the size of the schedule table. The question marks (???) occur in the schedule because the schedule is looking for schedule data, such as property set definitions, but none are attached to the objects.

This is what your finished drawing should look like.

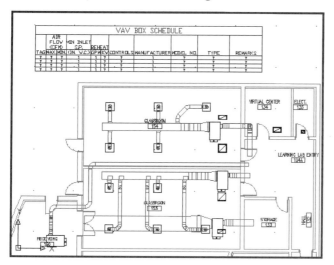

In this exercise you imported a predefined schedule table style into your drawing using Style Manager. You used Quick Select to select schedule all of the VAV boxes in your drawing and then placed the schedule table into your drawing. The next exercise takes you through the two different ways to attach schedule data to the objects, and also shows you how to add schedule data to the styles.

## EXERCISE 2: ADDING SCHEDULE DATA

In this exercise you correct the question marks (???) in the schedule table from the last exercise by establishing a relationship between the property set definitions that are being read by the schedule and the VAV boxes that you are scheduling. The VAV schedule you use in this lesson has been created by Autodesk and uses Property Set Definitions defined for both the VAV MvPart style and the individual VAV boxes.

 **Open:** *05 Adding Schedule Data.dwg*

### Manually Attach the Object Property Set Definitions to Individual VAV boxes

1. Attach Schedule Data to a VAV box:

   a. From the Documentation menu, click Schedule Data ➤ Attach/Edit Schedule Data.

   b. Select the top VAV box and press ENTER.

   c. In the Edit Schedule Data dialog box, click the Categorized tab and Click Add.

2. Review the attached property set definition:

   a. In the Add Property Sets dialog box, verify that the MvParts-VAVObjects property set definition is checked, and click OK.

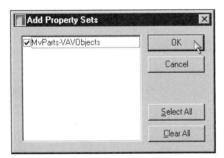

   b. In the Edit Schedule Data dialog box, change the property Tag to **VAV-124** and click OK.

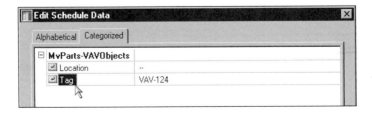

**Automatically Attach the Property Set Definitions to Individual VAV Boxes by Tagging**

3.  Add an object tag:

    a.  From the Mechanical menu, click Tags and Schedules ➤ HVAC Tags.

        DesignCenter opens to the available HVAC tags.

    b.  In the right panel of DesignCenter, scroll to the bottom of the tags and left-drag the VAV Tag3 into the drawing.

    c.  Select the VAV at the lower left of the drawing and pick a point to place the tag.

4.  Edit the Schedule Data:

    a.  Renumber the Tag to **VAV-123** and click OK.

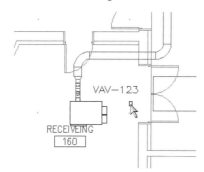

    b.  You are still in the tag routing, so select the VAV Box to the right and number the Tag **VAV121**, and then select the VAV Box above it and number the Tag **VAV-122**, and press ENTER.

    c.  Select the VAV Box that you added the property set definition to (VAV 123), right-click, and click Edit Schedule Data.

        Notice that the VAV number is read into the Tag.

        Some tags, such as the air terminal tag, use automatic numbering and number themselves incrementally.

        Tags provided with the software use whatever the current text style is for their attribute, therefore, make sure your office standard is current before you place the tags.

        The schedule still has many question marks in it. This is because the information the schedule is looking for exists in the property set definition for the style. Adding information to the VAV box style is very similar to manually attaching the property set definitions to the individual VAV Boxes.

        Click Cancel to exit the dialog box and end the command.

**Add Schedule Data to the VAV Box Styles**

5.   Edit the VAV box style:

   a.   Select the top VAV box, right-click, and click Edit MvPart Style.

   b.   On the General tab, Click Property Sets, and then click Add. Verify that the MvParts-VAVStyles property set definition is checked, and click OK.

   c.   Review the two types of property sets, manual and automatic.

   An ENTER key represents the manual property set and a lighting bolt represents the automatic property set.

6.   Change the property values:

   a.   Change the property CFM Max to **1900**, and the property CFM Min to **800**.

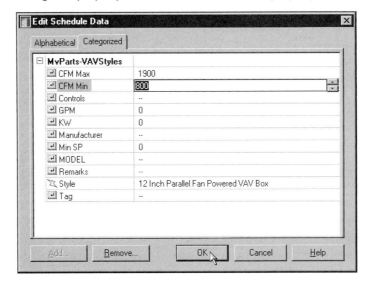

   b.   Click OK twice.

   In the drawing, the VAV boxes 122 and 124 have their Schedule Data filled out. Both of them have the same data because they are the same style.

| | AIR FLOW (CFM) | | MIN INLET S.P. | REHEAT | | | | | |
|---|---|---|---|---|---|---|---|---|---|
| TAG | MAX. | MIN. | (IN. W.C.) | GPM | KW | CONTROLS | MANUFACTURER | MODEL NO. |
| | ? | ? | ? | ? | ? | ? | ? | ? |
| VAV-122 | 1900 | 800 | 0 | 0 | 0 | -- | -- | -- |
| VAV-123 | ? | ? | ? | ? | ? | ? | ? | ? |
| VAV-124 | 1900 | 800 | 0 | 0 | 0 | -- | -- | -- |

VAV BOX SCHEDULE

c.  Repeat steps 5 and 6 to edit the two remaining VAV boxes styles by adding the MvParts-VAVStyles property set definition.

The schedule should be filled out.

This is what your finished drawing should look like.

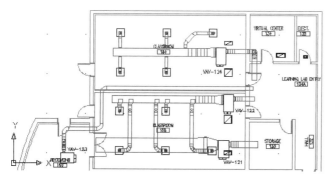

In this exercise you added schedule data to the VAV boxes in your drawing by attaching property set definitions to the VAV boxes. You did this using the Attach/Edit Schedule Table dialog box, as well as using object tags from DesignCenter. You then added information to the VAV box style in order to fill in the missing information from your schedule. In the next exercise, you customize the schedule table by adding and deleting columns.

### EXERCISE 3: ADDING A COLUMN TO THE SCHEDULE

In this exercise you add and delete columns in the schedule by editing the schedule table style. To add or delete columns to the schedule table the property must exist before it can be added into the schedule's style.

**Open:** *05 Adding Schedule Column.dwg*

**Add a Property to the MvParts-VAVObjects Property Set Definition**

1.  Access the formatting information for the schedule table:

a.  From the Documentation menu, click Schedule Data ➤ Property Set Definitions.

b.  In the left panel of Style Manager, double-click on MvParts-VAVObjects to open it for editing.

c.  Click the Definition tab and click Add to add a manual property. Enter **CoilLoad** for Name and click OK.

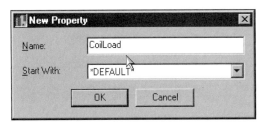

Start with the default because you may find an existing property that has the same type of information that you want to use.

You may add automatic properties by selecting the "Add Automatic" button to bring up a list of the automatic properties for all MvParts.

2.  Specify the formatting for the schedule table:

a.  Change the Type to Real and change the default to "0" so that you have something in the schedule to edit if needed.

You will be editing the schedule table in the next exercise.

b.  Change the format to Number - General.

This is the schedule data format that you saw in Style Manager.

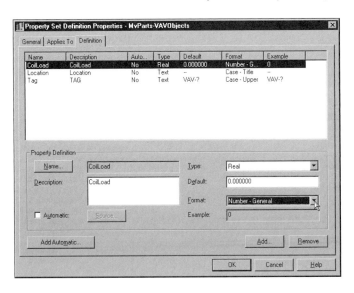

c.  Click OK twice.

Now that the Property has been created, you can place it in the schedule as a column.

### Add and Delete Columns in the Schedule Table

3.  View the property sets in the drawing:

    a.  Select the schedule table in your drawing, right-click, and click Edit Table Style.

    b.  On the Columns tab, click Add Column.

    This brings up a list of all of the properties defined in property sets in the drawing.

    c.  Find MvParts-VAVObjects CoilLoad and highlight it. On the right, verify that the heading is CoilLoad.

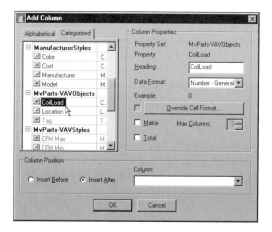

4.  Delete a column:

    a.  Select Insert After and use the column drop down list to select MvParts-VAVStyles: KW and then click OK.

    b.  On the Columns tab, scroll over and select the heading Model No., click Delete, and in the Remove Columns/Headers dialog box, click OK.

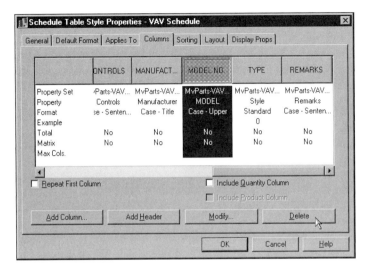

c.  Click OK.

Zoom into the schedule to see that the column CoilLoad is added to the schedule. The Model number column is now gone.

| VAV BOX SCHEDULE | | | | |
|---|---|---|---|---|
| REHEAT | | | | |
| GPM | KW | CoilLoad | CONTROLS | MANUFACTURER |
| 0 | 0 | 0 | -- | -- |
| 0 | 0 | 0 | -- | -- |
| 0 | 0 | 0 | -- | -- |
| 0 | 0 | 0 | -- | -- |

 **Note:** You can also move columns in the schedule by left-dragging them within the layout using the options on the Column tab. The Column tab is also where you can add headers on top of two or more columns by holding down the CTRL key while you pick the headers in order to select multiple headers, and then click the add header button.

This is what your finished drawing should look like.

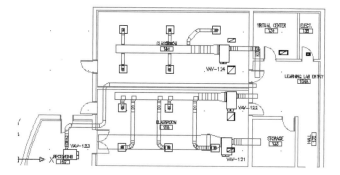

In this exercise you changed the formatting of the schedule table by adding property definitions to the schedule table style. In the next exercise you will be editing the information contained in your schedule.

### EXERCISE 4: EDITING SCHEDULE DATA

In this exercise you will review some of the ways to edit the data in the schedule using the pop-up tools available from the shortcut menu. All of these tools are also available from the Documentation menu under Schedule Tables.

 **Open:** *05 Editing Schedule Data.dwg*

### Editing Cells Directly in the Schedule Table

Editing cells enables you to directly access the schedule data. There can be up to four different types of schedule data displayed by any schedule. These four types of schedule data are: Manual and Automatic properties for the *style* of the scheduled objects and Manual and Automatic properties for the *individual objects.*

You can directly edit any objects that contain manual properties. You can directly edit style manual properties; however, this brings up some warnings.

Automatic properties belonging to the style or object *cannot* be edited, as the value for these comes directly from the object.

### Edit a Manual Property for a Style

1.   Change a table cell's value:

a.  Zoom into the schedule table, select the schedule table, right-click, and
    click Edit Table Cell.

b.  Click the "–" under Manufacturer for VAV 123. The alert tells you that this
    belongs to the style; click Yes.

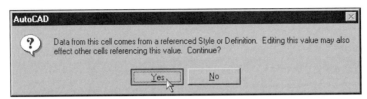

c.  In the Edit Schedule Property dialog box, enter **McQuay** for Value and
    then click OK.

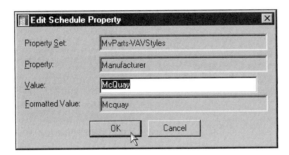

Because there is only one VAV of this style, only one cell fills out. Press
ENTER to end the command.

2.  Copy a VAV Box:

a.  Zoom out and select VAV Box number 123 and its associated object tag,
    right-click, and click Copy.

b.  Click in the upper right room, right-click, and click Paste.

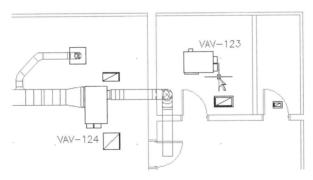

The pasted VAV Box has the same style so it contains the same information. The schedule data for VAV Box number 123 is also copied. You now have two VAV Box number 123's in the drawing.

c.  Select the schedule table, right-click, and click Selection ➤ Show. In the schedule select the second VAV 123.

Notice that the matching VAV is highlighted in the drawing. Press ENTER to end the command.

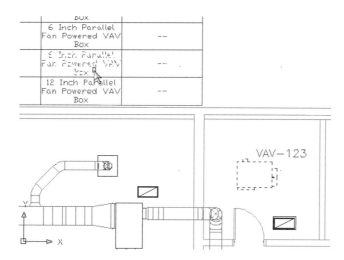

## Edit a Manual Property on an Object

3.  Edit a schedule table cell:

a.  Select the schedule table, right-click, and click Edit Table Cell.

b.  In the schedule table, select the second VAV Box, number 123.

c.  In the Edit Schedule Property dialog box, enter **VAV-125** for Value and then click OK.

The VAV Box's new tag number is updated in the schedule table, and the tag that reads from the same property set definition information is also updated. You can change the values in the schedule table directly because the property is attached to the individual object. Press ENTER to end the command.

## Edit an Entire Column

4.  Change a column in the schedule table:

a.  Select the schedule table, right-click, and click Edit Table Cell.

b. Pick on the line between the first two zeros in the coil WPD FT WC column to open the Edit Schedule Data dialog box that shows all of the property sets assigned.

| COIL | | |
|---|---|---|
| Load MBH | GPM | WPD FT WC |
| 0.0 | 0.0 | 0.0 |
| 0.0 | 0.0 | 0.0 |

c. Click the Categorized tab and change CoilWPD to 1, and then click OK.

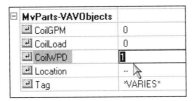

The entire column changes to 1. Press ENTER to end the command.

### Edit the Schedule Data

5. Change an individual object's properties:

a. Select VAV Box number 122 and its associated object tag, right-click, and click Edit Schedule Data.

b. The Edit Schedule Data dialog box opens showing the attached property sets.

c. Change CoilGPM to **5.7** and click OK.

The schedule is updated with this information

You can use any of these tools to edit the information that is held in the schedule. Keeping in mind that you may want to have your schedule in a different sheet than your model files, some of these methods may not work directly. For example, edit table cell does not work when the schedule is pointing at VAV boxes in an external reference file (xref). The schedule reads the information, but you do not have direct access to it.

You can keep the same schedule *style* in your model sheet, and then xref your model into a schedule sheet for plotting. If you need to modify any schedule data, you can place a temporary schedule in the offending model file to track your changes, make your edits, and then erase the temporary schedule. The next time you open your schedule's drawing, you can update the schedules and the new information will be read.

**Caution:** If you have selected Automatic Update on the Add Schedule Table dialog box when you placed the schedule, and you use the 'overlay' method of externally referencing, make sure that your model files where the schedule information is held is never more than one link back from where the schedules are held. The schedule reads the property set definitions attached to objects, but that information has to be available for the schedule to read. "Problem schedules" can occur where the schedule values all disappear because the overlay cuts off the schedule from the objects.

This is what your finished drawing should look like.

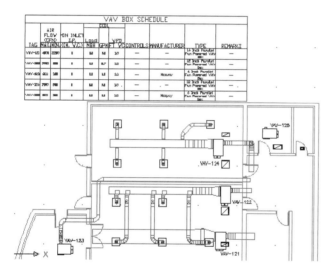

In this exercise you changed the information contained in the schedule table by editing table cells and columns in the schedule table. In the next exercise you will edit the schedule table style.

## EXERCISE 5: EDITING THE SCHEDULE TABLE STYLE

In this exercise you will edit the schedule table style by changing the text appearance and the line colors for the cell borders. The schedule table style defines the format and appearance of the schedule table.

**Open:** *05 Fine Tuning Schedules.dwg*

**Edit the Schedule Title Format**

1. Access the formatting information for the schedule table:

   a. Zoom into the schedule table; select the schedule table, right-click, and click Edit Table Style.

   b. Click the Layout tab and enter **VAV TERMINAL BOX SCHEDULE** for the Table Title.

   c. For Title, click Override Cell Format.

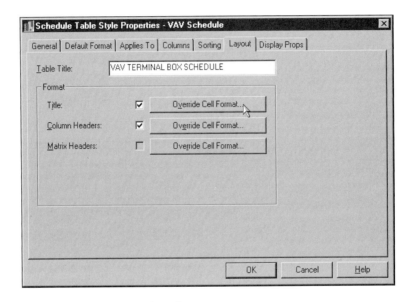

2. Change the format for the title cell:

   a. In the Cell Format Override dialog box, select Arial-Bold for Style.

   b. Enter **3/16"** for Height and click OK.

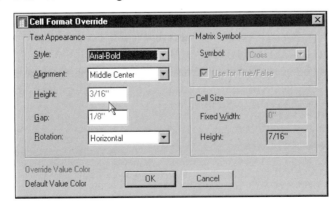

   c. For Column Headers, click Override Cell Format.

**Edit the Display Properties of a Column**

3. Change the format for the column header:

    a. In the Cell Format Override dialog box, select Arial-Bold for Style.

    b. Enter 1/8" for Gap and click OK.

    c. Click the Display Props tab and under Property Source, select Schedule Table Style.

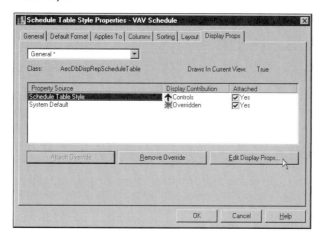

4. Change the color of the columns:

    a. Click Attach Override and then click Edit Display Props.

    b. On the Layer/Color/Linetype tab, change the color for the Outer Frame component to Red.

       To change the color of the component, under Color, click ByBlock, and in the Select a Color dialog box, click on a color square or enter a color number and then click OK.

    c. Change the color for the Major Row Lines component to 10.

    d. Change the color for the Minor Row Lines component to 94 and click OK twice.

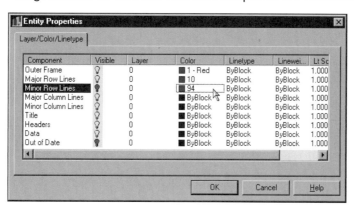

The schedule table reflects the formatting and color changes that you made; you may have to regenerate your drawing to view these changes by entering **REA** at the command line.

This is what your finished drawing should look like.

In this exercise you changed the information contained in the schedule table style by editing the display properties.

In Lesson 3, "Placing Schedules," you completed five exercises: Creating a Schedule, Adding Schedule Data, Adding a Column to a Schedule, Editing Schedule Data, and Editing the Schedule Table Style. You have now successfully scheduled the VAV boxes in your drawing, changed the VAV box information in your drawing and in your schedule, edited the schedule table to reflect the columns that you wanted to show, and edited the schedule table style in order to format the schedule table.

# KEY CONCEPTS: REVIEWING YOUR DESIGNS

Disconnect Markers alert you to connectors or connections that are not valid because the nodes are misaligned, or the systems of the two pieces do not match.

Interference Detection appears when a Building Systems object such as a duct or pipe fitting crosses through another Building Systems object or structural member.

Structural members are Architectural Desktop objects such as beams and columns.

A 2D section object enables you to view where the different Building Systems objects are in relation to one another.

Schedules enable you to produce a detailed list of all of the objects in your drawing.

There are three different style definitions needed to create a schedule table: Schedule Data Formats, Property Set Definitions, and Schedule Table Styles.

Property Set Definitions are a group of properties that hold schedule information.

Properties come in two types: Automatic and Manual.

Automatic properties read values such as style name or dimensions from the AEC objects. Once placed in a schedule you do not have direct control over these properties through the schedule.

Manual properties are filled in by you, and may be edited directly in the schedule table.

Schedule Table Styles hold the organization of the schedule table; for example, the order of the columns.

A Schedule Tag is created with an attribute that looks at the schedule data attached to the object and reads the properties out of the schedule data, and then reports that data as its own attributes.

CHAPTER 6

# Communicating Your Design

After you read this chapter, you should understand the following:

- how to control the annotation of objects in your drawing
- how to annotate your mechanical plan
- how to share your design with other engineers
- how to enable others to view your mechanical plan using live enablers and proxy graphics
- how to plot your mechanical drawing
- how to save in Drawing Web Format (DWF)

## TOOLS USED TO COMMUNICATE YOUR DESIGN

This chapter covers different drawing sharing techniques in order to hand off your mechanical design to building contractors, suppliers, architects, electrical engineers, and fellow mechanical engineers and draftsmen. These communication tools include posting and plotting your drawing, using live enablers for others who do not own Autodesk Building Systems, and saving your drawing in Drawing Web Format (DWF) so that you can post your drawing to an Internet site.

### LESSON OBJECTIVES

This chapter contains three lessons. The lessons are broken down into one or more exercises. Most exercises have a corresponding drawing file that is located on the CD-ROM included with this book.

Lessons in this chapter include:

- **Lesson 1: Annotating Objects in Your Drawing**
  You annotate the objects in your drawing in several ways, including assigning flow direction arrows, applying properties to ducts such as insulation and lining, and adding rise and drop symbols.

- **Lesson 2: Annotating Your Mechanical Plan**
  You will add text symbols, system labels, and Autodesk Architectural Desktop annotation symbols such as section marks and detail references, to clarify your mechanical design.

- **Lesson 3: Distributing Your Mechanical Design**
  You will learn about the tools available to you so that you can share and distribute your mechanical design to others.

## LESSON 1: ANNOTATING OBJECTS IN YOUR DRAWING

Objects contain display information that Building Systems reads and then displays on your screen. This display information can be accessed through styles, display properties, and as separate commands. This lesson explains several different ways of using the display information that is built into objects.

This lesson contains six exercises. Each exercise has a drawing file associated with it. When you have completed the exercises in this lesson you will have used both Building Systems and Architectural Desktop annotation tools to emphasize the important aspects of your mechanical design. At the end of each exercise there is a screen shot of what your finished drawing should look like.

This lesson contains the following exercises:

- Placing Flow Direction Arrows
- Viewing Duct Insulation and Lining
- Adding Rise Drop Symbols in Ducts
- Adding Vanes to Elbows

- Adding Flow Arrows to Air Terminals
- Changing the Display of an MvPart

## EXERCISE 1: PLACING FLOW DIRECTION ARROWS

There are three different methods for placing flow arrows in your drawing. You can create a label style that has a custom arrow block assigned to it and then label the duct with a flow arrow. You can apply an override to a duct's display properties and then assign a custom block to the current display representation. You can also use the flow direction arrows. In the following exercise you place flow direction arrows using the Set Flow and Show Flow commands. These are available from the MEP Common pulldown menu by clicking Utilities ➤ Set Flow, and Utilities ➤ Show Flow. You must specify the flow direction using the Set Flow command before the Show Flow command can display the flow arrows. You can use the annotation setting on the Scale tab of the Drawing Setup dialog box to specify the size of the flow arrows that are displayed, but keep in mind that this setting also affects other annotations, such as labels, if they exist in the drawing.

**Open:** *06 Flow Direction Arrows.dwg*

### Change the Annotation Settings

1.  Verify the annotation plot size:

    a.  From the Desktop menu, click Drawing Setup.

    b.  In the Drawing Setup dialog box, click the Scale tab.

    c.  Change the Annotation plot size to **1/8"** and then click OK.

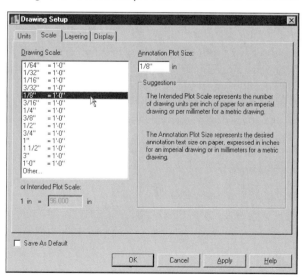

## Show Flow Arrows

2.  Show flow arrows for a selected duct run:

    a.   From the MEP Common menu, click Utilities ➤ Set Flow and pick the horizontal duct on the left.

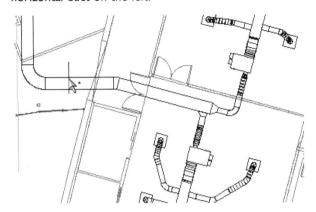

    b.   On the command line, enter **n** to place the flow direction arrows pointing to the right.

       The direction of the arrow is dependent on the direction the duct was originally placed in the drawing. In this case, it just happens to be in the correct direction.

    c.   From the MEP Common menu, click Utilities ➤ Show Flow, and enter **y** at the command line to display the flow direction arrows.

       The arrows are shown for the run of duct that you selected. The arrows stop at a branch and you must reissue the command for the next run of ducts.

## Create a Flow Direction Arrow out of a Custom Block

While the standard flow arrows are by far the easiest way to show flow direction on a duct or pipe, you may find that the standard flow arrows lack the control you need regarding size, number, and placement. You can create your own custom flow arrows by adding a standard AutoCAD block to the display representation of the duct. In this next part of this exercise you will assign a block to the display representation of the duct. There is a block in the drawing created for you to use in this exercise. The block that has been created for you is a standard AutoCAD block called 'CustomFlowArrow'.

3. Access the entity display properties of the duct:

    a. Select the main trunk of the duct coming out of the upper VAV box, right-click, and click Entity Display.

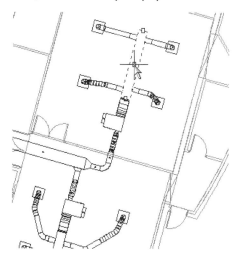

    b. Click the Display Props tab, and under Property Source select Duct and then click Attach Override.

    c. Click Edit Display Props, and in the Entity Properties dialog box click the Other tab.

4. Add a custom block to represent a flow arrow:

    a. On the Other tab under Custom Block Display, click Add.

    b. In the Custom Block dialog box, click Select Block and then scroll down in the Select A Block dialog box to select CustomFlowArrow.

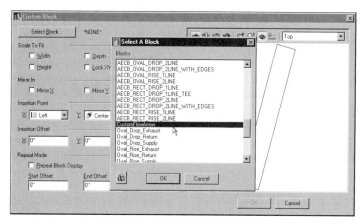

    c. Click OK to return to the Custom Block dialog box.

5.  Define the placement of the custom block flow arrow:

    a.  Under Insertion Point, change the X to center, and verify that Y and Z are already set to center.

    b.  If the arrow is backwards for this duct, then click the Mirror X.

    There are many X, Y, and Z controls. To understand these, think of a duct drawn from left to right in the World UCS where X is along the World UCS X *axis*, Y along the World UCS Y *axis* and so on.

    c.  Under Insertion Offset, enter 12" for X and click OK.

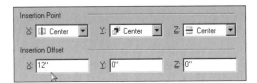

Changing the insertion offset moves the custom flow block 12" along the duct. This avoids overlap with the show flow arrows created in the first part of this exercise. You can use the Y offset to move the arrow to one side of the duct or the other.

6.  Change the color of the custom block flow arrow:

    a.  Click the Layer/Color/Linetype tab and notice that you now have display control over the custom block that you added.

    b.  Change the color of the custom flow arrow to blue.

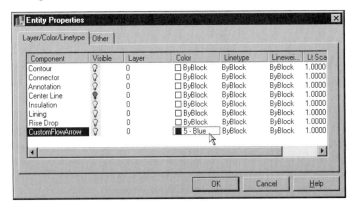

    c.  Click OK three times to return to drawing.

The benefit of creating a flow arrow out of a custom block is the ability to override the display representation so that you have more control over the size and color of the flow arrow. You can change the size of the CustomFlowArrow block that you inserted by entering **refedit** at the command line. Refedit redefines the block in the drawing, and changes all instances of that block that have been added to the display representation of the ducts.

The benefit of using the Show Flow Arrow command is that it is not display dependent, therefore, if it is turned on, all viewports are displayed regardless of the display configuration that is assigned to the viewport.

You can click the layout tabs to view the flow arrows in the different display configurations. Notice that only the layouts that use the 2-line representation of the duct display the custom block; however, all of the layouts display the flow arrows placed using the Show Flow command. If you wanted to have the custom block appear in the 1-line representation, you would have to repeat the process for the 1-line representation of the duct.

This exercise provided two examples of how to show the flow direction on a duct or pipe; by adding a custom block and by using the Show Flow command. You can add a custom block to any display representation for ducts, pipes, and fittings, but not to MvParts (multi-view parts).

### EXERCISE 2: VIEWING DUCT INSULATION AND LINING

Insulation or lining may be applied to the ducts as they are drawn, or after they are placed in the drawing. The settings for these two actions are different. If you want Building Systems to add lining or insulation as you draw ducts and fittings, you need to specify this in the Duct Layout Preferences on the Duct tab. If you want to add insulation or lining to an individual duct after it is drawn, you need to access the duct or fitting's properties page and go to the Lining and Insulation tab. Once the insulation or lining is applied, you can control the display with the system style display properties.

**Open:** *06 Lining insulation and flex.dwg*

### Adding Insulated Duct

1.  Specify the duct's insulation thickness:

    a.  From the Mechanical menu, click Duct ➤ Preferences.

    b.  Click the Ducts tab, select Apply Insulation, and enter **1.5** for Thickness.

    c.  Click OK.

2. Add a duct to the drawing:

   a. From the Mechanical menu, click Duct ➤ Add.

   b. In the Add Ducts dialog box, select Supply for System.

   c. Click in the drawing to make it active and specify a start point at the left side of the coil. Drag your cursor to the left and enter **18"** at the command line to specify the length of the duct.

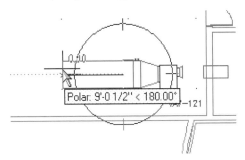

   An 18" piece of duct is added.

3. Change the duct height and width:

   a. In the Add Ducts dialog box, enter **40** for the Width and enter **20** for the Height.

   b. Pick a point about 2/3 the remaining distance to the left wall.

   c. In the Add Duct dialog box, change the Shape to Round and enter **16"** for the Diameter.

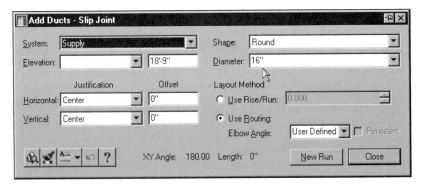

   d. Pick a second point toward the wall and then a pick a third point below the break mark, and select close in the add ducts dialog box.

All the ducts and fittings are shown with insulation.

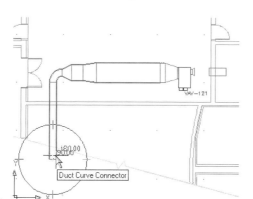

## Add Flexible Duct without Insulation and Change the way that Flexible Ducts Display

4.  Remove insulation from duct preferences:

    a.  From the Mechanical menu, click Duct ➤ Preferences.

    b.  Click the Ducts tab and deselect Apply Insulation.

    c.  Click the Flex Ducts tab, and under 1-Line Annotation change the Graphics to Curve pattern and enter **8"** for Pitch.

    d.  Under 2-Line Annotation, change the Graphics to Vertical pattern, enter **8"** for Pitch, and then click OK.

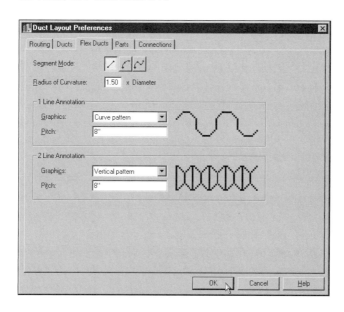

5.  Connect the VAV Box with flexible duct:

    a.  From the Mechanical menu, click Flex Duct ➤ Add.

    b.  Select Supply for System and verify that the Diameter is set to 20".

    c.  Click in the drawing to make it active and draw a section of flex duct between the fitting on the right side of VAV 121 and the left end of the 20" round duct at the right.

    The flex duct takes on the graphics specified in the Duct Preferences dialog box.

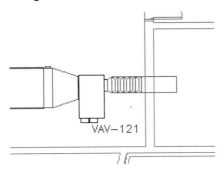

6.  Change the display of a flexible duct:

    a.  Select the segment of flexible duct at the right of VAV 124.

    b.  Right-click, and click Flex Duct Properties.

    c.  Click the Flex Graphics tab, and under 2-Line Annotation change Graphics to Vertical pattern, enter **8"** for Pitch, and then click OK.

7.  Add insulation to a duct:

    a.  Select the main trunk of the branch from VAV 124.

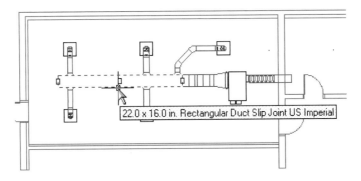

   b.   Right-click, and click Duct Properties.

   c.   On the Lining and Insulation tab, select Apply Insulation and enter **1.5"** for Thickness, and then click OK.

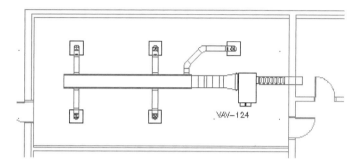

Only the single instance of the flexible duct is modified, because the lining only applies to the selected duct. You can select multiple ducts if you want to change their properties in order to add insulation. However, you would then also have to select all of the fittings to add the insulation to as well. You can control the display of the insulation separately from the duct. To apply the same display of insulation to the entire system, you can specify display properties within the system definition.

### Edit the System Definition to Change the Display Properties

By default, the duct and the insulation are the same color and linetype. The insulation is a subobject of the duct. You can change how the insulation is depicted or displayed by changing the display properties of the duct using Duct Properties, or you can change how all of the ducts in the system display their lining or insulation. The last bit of this exercise illustrates how to change the display representation for the supply system in order to show the insulation as magenta. Although the layer of the insulation is not changing, it will still be part of the duct, and assigned to the same layer as the duct.

8.   Access the HVAC system definition:

   a.   From the Mechanical menu, click Mechanical Systems ➤ HVAC System Definition.

   b.   In Style Manager, right-click on the Supply system definition and click Edit.

   c.   Click the Display Props tab, and under Property Source select Duct System Definition.

9. Edit the HVAC system definition:

   a.   Click Attach Override, and then click Edit Display Props.

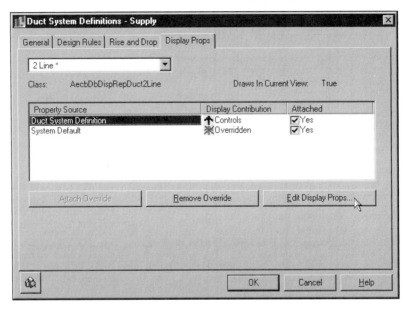

   b.   Change insulation color to magenta and click OK.

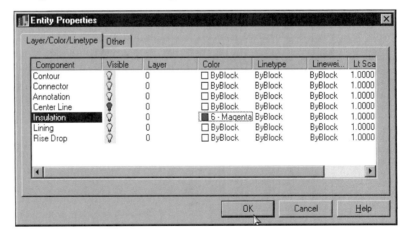

   c.   Click OK three more times to return to the drawing.

      The duct insulation now displays in a magenta color. If you need layer control over the lining or insulation, you could create a new layer called M-Duct-Insl and change the layer in the display properties, and change the color to ByLayer.

### EXERCISE 3: ADDING RISE/DROP SYMBOLS IN DUCTS

All vertical segments of duct are always assigned the drop symbol by default. If you use the Set Flow command on the branch, then the rise/drop symbol is displayed correctly.

By default the behavior of the rise/drop symbol always places the drop symbol in plan when the duct is placed. After you use the Set Flow command, the rise/drop symbol appears correctly. The standard rise symbol displays as a solid line and the drop symbol displays as a dashed line.

A duct's rise and drop symbol is controlled by a Rise/Drop style accessible in Style Manager. The Rise/Drop style is assigned to the HVAC System Definition. If you want to modify the way that the rise/drop symbol is displayed, then you can modify the Rise/Drop style using Style Manager.

**Open:** *06 Rise and Drop.dwg*

**Set Flow to Show Correct Symbols**

1.  Set flow for the make-up air duct:

    a.  From the MEP Common menu, select Utilities ➤ Set Flow.

    b.  In the left viewport, select any duct in the makeup air run (below the Air Handler, be careful not to select a fitting).

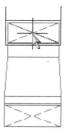

    c.  At the command line, enter **n** to leave the flow direction going into the economizer.

        The symbols at each end where the duct travels away from you stays dashed to indicate a drop in the duct run.

2.  Set flow for the supply duct:

    a.  At the command line, enter **setflow** and then press ENTER.

    b.  Select any segment of the supply duct, but be careful not to select a fitting.

    c.  At the command line, enter **n** to leave the flow direction as is.

The cross symbol where the duct turns up changes to solid, indicating a rise in the duct.

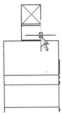

3. View the Rise/Drop styles:

    a. From the Mechanical menu, click Mechanical Systems ➤ HVAC System Definition.

    b. In the right panel of Style Manager, select Make-up Air, right-click, and click Edit.

    c. Click the Rise and Drop tab and view the Rise/Drop styles that are available to you, and then click OK twice to end the command.

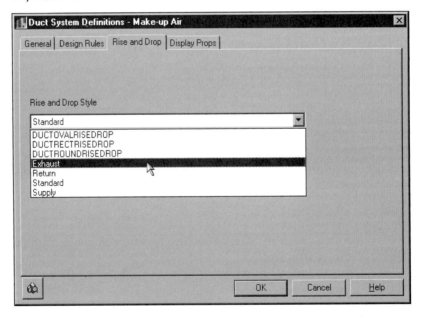

Each of the Rise/Drop styles use standard ACAD Blocks to represent the rise or drop for the different representations I-Line, 2-Line, etc., as well as for each of the different possible duct shapes.

### EXERCISE 4: ADDING VANES TO ELBOWS

Vanes are added to elbow fittings. When you add vanes to an elbow, the vane gets added to the fitting style and all elbows of that size display the vanes as well. Therefore, if you would like to display vanes on the supply system but not the return system, you have to edit each elbow individually on the return system and use the Layer/Color/Linetype tab of the Entity Properties dialog box to turn off the display of the vanes. The drawing used for this exercise is simulated to get the point across. Watch the command line as you are prompted for several things as you complete the Add Vanes command.

**Open:** *06 Elbow Vanes.dwg*

**Add Vanes to Elbows**

1.  Access the Add Vanes command:

    a.  From the Mechanical menu, click Duct Fittings ➤ Add Vanes.

    b.  At the left-hand side of the drawing, select the Supply elbow.

    c.  Use end OSNAPs to select the upper-right inside corner of the elbow as the start point, and the lower-left corner as the end point.

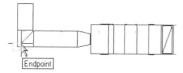

    If you want asymmetrical vanes, you can place some nodes to define your desired pick points prior to adding vanes. If you do this, the vanes will be distributed along the line that you establish with the first two pick points.

2.  Define the display of the vanes:

    a.  At the command line, enter **18"** for the radius and then press ENTER.

    b.  Enter **60** to define the angle of the arc and then press ENTER.

    c.  Press ENTER to accept 4 as the default number of vanes, enter **n** to erase the path, and then press ENTER to end the command.

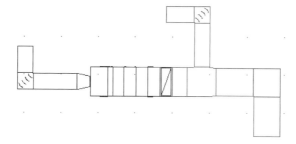

**Use Display Override to Turn off Vanes for a Single Elbow**

3. Access Entity Display:

a. At the right-hand side of the drawing, select the return elbow that displays vanes, right-click, and click Entity Display.

b. Click the Display Props tab, and under Property source select Duct Fitting and click Attach Override.

c. Click Edit Display Props.

4. Turn off Vanes:

a. Click the Layer/Color/Linetype tab and select Annotation.

b. Click the light bulb to turn off the annotation for the return elbow.

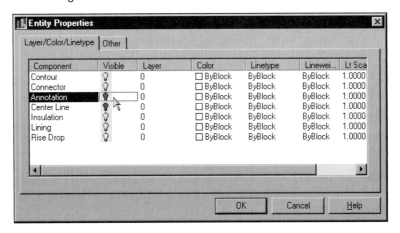

c. Click OK twice to return to the drawing.

The vanes on the return elbow no longer display.

 **Tip:** The following information outlines a limited workaround if you are not doing quantity takeoff schedules, and do not use transitional mitered elbows in your drawing.

5. Place your supply system using the rectangular mitered elbow as default.

Placing your return system using the rectangular transitional mitered elbow as default works for turning the vanes on for ducts of the same size.

6. Make a copy of the rectangular mitered elbow using Catalog Editor and rename the elbow something unique, for example, *rectangular mitered elbow return*.

### EXERCISE 5: ADDING FLOW ARROWS TO AIR TERMINALS

Air Terminal Flow Arrows are part of any of the air terminal MvParts. If the parts type is Air_Terminal, then it has an extra tab on the properties page called the Flow Annotation tab. This drawing contains a custom arrow created for you that you will use as a flow arrow for the Air Diffusers. While you can only select one Air Terminal MvPart to modify, the MvPart Properties allows multiple parts in the same selection set. In this exercise you select all of the Air Diffusers in the drawing, and then adjust the properties for all four-air diffusers at once, then edit the style of the air diffuser to adjust the color of the attached block. One of the air diffusers is 12 x 24 to show that the insertion point is based on the edge of the air diffuser, not the center.

**Open:** *06 Air Terminal Arrows.dwg*

### Assign the Custom Block to the Air Diffusers

1.  Select the four air diffusers:

    a.  From the Tools menu, click Quick Select.

    b.  For Object Type select Multi-View Part; for Properties, scroll down and select Type.

    c.  Enter **Air_Terminal** for Value and then click OK.

    The four air diffusers are highlighted.

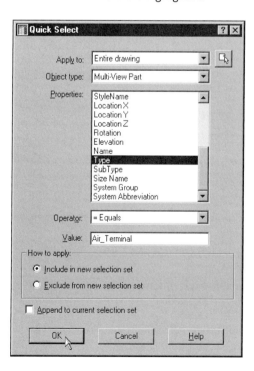

2.  Specify the flow direction to display on the four air diffusers:

    a.  Right-click and click MvPart Properties.

    b.  Click the Flow Annotation tab and select the up, down, left, and right arrows.

    c.  Click Select Block, and in the Select A Block dialog box select 06CustomDiffuserArrow and then click OK to return to the Flow Annotation tab.

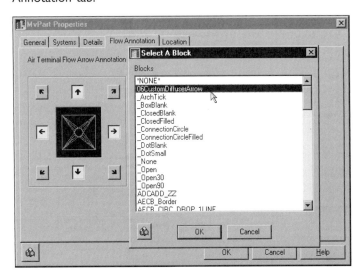

The 06CustomDiffuserArrow is a standard AutoCAD block that was created for you to use in this exercise.

3.  Specify the flow direction arrow placement:

    a.  Select Use the Same Block for All Arrow Locations.

 **Note:** If you have a rectangular diffuser you need to create a block that uses a different distance from the insertion point to clear the block.

    b.  Verify that the Rotation Angle for This Block and the Offset from Edge of Block is set to 0.

    c.  Verify that Use Annotation Scale is deselected and click OK.

### Adjust the Style of the Air Diffuser to Change the Color of the Arrows

4.  Edit the style of the Air Diffuser:

    a.  Select one of the air diffusers, right-click, and click Edit MvPart Style.

    b.  Click the Display Props tab, and leave System Default selected under Property Source.

    c.  Click Edit Display Props.

5. Change the color of the arrows:

    a. In the Layer/Color/Linetype dialog box, select Annotation.

    b. Change the Annotation color to 134 and then click OK two times.

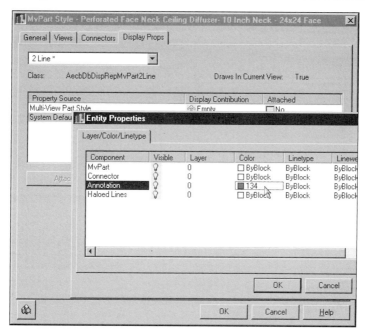

    c. Enter **regen** at the command line and ENTER to display the new color.

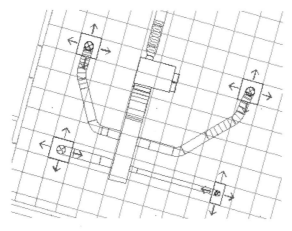

The only MvPart that uses the Annotation subcomponent is Air Terminals; therefore, changing the system default does not affect the other MvParts in the drawing, because the other MvParts do not have Annotation as a subcomponent.

218

## EXERCISE 6: CHANGING THE DISPLAY OF AN MVPART

In this exercise, you assign a block with an X in it to represent the air diffusers. You do this rather than using the view block assigned to the part that comes from the catalog.

In this drawing, two layouts have been modified in order to emphasize that different blocks are used to represent the MvPart (multi-view part) in different display configurations.

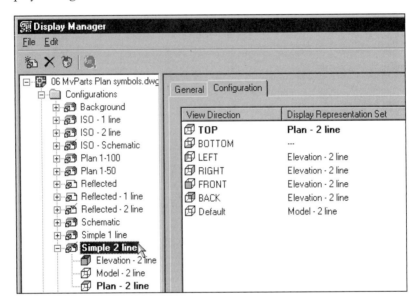

The drawing opens to the layout Plan 2 line which has the following Display settings: Display Configuration = Simple 2 line, Display Set = Plan 2 line, which uses the 2 line representations of MvParts.

 **Open:** *06 MvParts Plan symbols.dwg*

### Make the ISO-Schematic Viewport Current

1. Review the schematic display settings:
   a. From the Desktop menu, click Display Manager.
   b. Expand the Configurations folder, and in the left panel, select ISO-Schematic.
   c. In the right panel, select the Configuration tab.

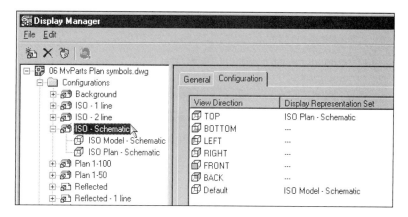

Display Configurations are assigned to a viewport. The Display Configuration assigns Display sets to possible view directions. Cancel out of the Display Manager and return to the drawing.

d.  Pick the layout tab Plan 2 Line-RCP at the bottom of the screen. As setup in the templates, the Plan 2 Line-RCP layout tab uses the same Display Set for all view directions. We have added a couple of viewports to this layout and assigned the Simple 2-line display configuration to these viewports to help illustrate how the display set is used by the MvParts.

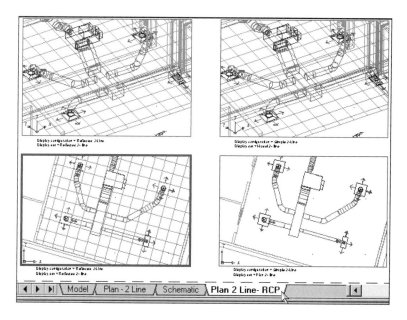

**Edit the Diffuser Style to Assign a Different Block to the 2-Line Display**

2. View the Two Line Top display representation:

   a. In the lower-left viewport, select the bottom-left square diffuser, right-click, and click Edit MvPart Style.

      It may be easier to select the diffuser if you zoom in and select the diffuser in the center.

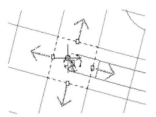

   b. Click the Views tab.

      You can use the Views tab to add a block and assign it to a display representation, or you can modify one of the existing definitions.

   c. Scroll down and select item #9 with the view name Two Line-Top.

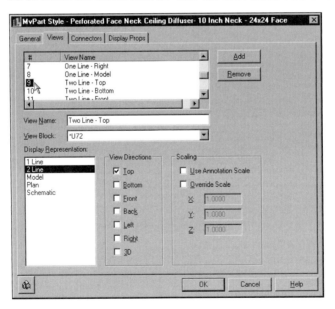

      This shows you that the anonymous AutoCAD view block *U72 is assigned to this view of the MvPart. This is typical for MvParts. The original block is held in the .DWG file that is part of the catalog. When the MvPart is placed in the drawing, the .XML file stores which blocks make up this part. These AutoCAD blocks are placed in the current drawing, and given anonymous names. If you want to change how this part behaves or displays for all of your drawings, then you have to modify two parts of the catalog. Both the .DWG and the .XML need to be modified. For more information, see Chapter 7, "Creating Custom Content."

3. Assign a different block to the Two Line Top display representation:

    a. In the MvPart Style dialog box, use the View block drop-down list to change the view block to 06SupplyDiffuserPlan24x24, and then click OK.

    b. Notice that all representations of this MvPart, including the Two Line Top view, have been changed to include an x representation of the diffuser.

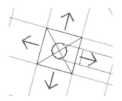

 **Note:** The ISO Schematic display configurations use a similar block.

    c. Zoom back out to see that the smaller diffuser has not changed because it is a different style of MvPart.

## LESSON 2: ANNOTATING YOUR MECHANICAL PLAN

There are many ways to annotate your mechanical plan. Some annotations, such as the system and size of ducts, may be Building Systems labels. Other annotations, such as title marks, section marks, and detail references, may be Architectural Desktop labels. This lesson looks at both of these products and all of the labeling features that you will need to use for your drawings. In either case, the setting on the Scale tab of the Drawing Setup dialog box determines the size that your label displays.

The labels included in Building Systems are used to provide annotations to your drawing. A label is a style-based object, that when placed in the drawing has an anchored relationship with the object to which it is assigned. It is through this anchored relationship that the label is able to display information about the object. The information shown by the label is held in the label style. The style determines if the label will display the system of the pipe or the size. The label could also be an AutoCAD block that you created and assigned to the label to be used as a flow arrow. A style also has the variables that control the appearance of the text. Any standard AutoCAD text style may be assigned to a label style. If the AutoCAD text style is assigned a height, then the height is assigned to the label. If the AutoCAD text style is assigned 0 as the height, then the label text is sized according to the proportions set in the Drawing Setup dialog box on the Scale tab.

This lesson contains two exercises. Each exercise has a drawing file associated with it. When you complete the exercises in this lesson you will have used both Building Systems and Architectural Desktop labeling tools to emphasize the important aspects of your mechanical design.

This lesson contains the following exercises:

- Using Building Systems Labels
- Using Architectural Desktop Labels

### EXERCISE 1: USING BUILDING SYSTEMS LABELS

In this exercise you will apply Building Systems labels and utilize the layer key override to assign different layers to the plumbing and duct labels. Labels are a Building Systems feature that place a piece of text on a pipe, duct, or MvPart. The label looks at the object being labeled to display a specific set of information. There are three label styles available for you to use in the drawings started from the template: name, standard, and system label.

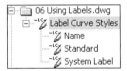

You will be using the system label that reads the abbreviation out of the system definition and displays it in the drawing.

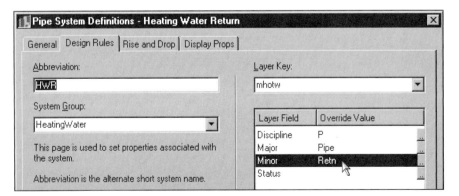

This screen caption illustrates a couple of things. The label looks in the system definition for the abbreviation that it uses as you place a label on a pipe. The system definition shown here also illustrates the layer key override used by the pipes (the drawing was set up this way to accommodate the supply and return pipes on different layers). You use a similar layer key override manually when you place the labels for the pipes. The label feature is very command-line-oriented, so watch the command line closely. The prompts are explained in the steps of the exercise.

 **Open:** *06 Using Labels.dwg*

### Place Duct Labels

1.  Select the duct to label:

    a.  From the MEP Common menu, click Labels ➤ Add Labels.

    The command line asks for a curve. A curve is any linear element; for example, a schematic line, duct, or pipe.

    b.  Select the upper main duct trunk.

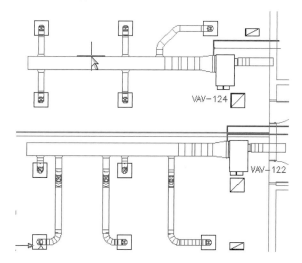

    c.  At the command line press ENTER to accept the standard style and then enter **m** to create manual spacing.

    The other options are space evenly or repeat. Space evenly places a set number of labels evenly along the length of the duct. The repeat option places a defined amount of labels at a set distance.

2.  Label the duct:

    a.  At the command line, enter **1** to place just one label along the duct.

    b.  Select a point on the duct for the Node Position.

    The relationship between the duct and the label is an anchor; this anchor can be compared to the anchor that is used with Schematic Symbols and Lines. The "curve" selected becomes the base restraint, and the "Node" is the point of the anchor.

### Copy a Label to Another Duct

When you copy a label from one duct to another you need to establish the anchor to the duct that you are pasting to.

3.  Select the label to copy:

    a.  Select the label that you just placed, right-click, and click Copy Selection.

    b.  The command line prompts you to specify a base point or displacement; select another place on the upper main duct trunk.

    c.  Select the lower duct trunk as the point of displacement.

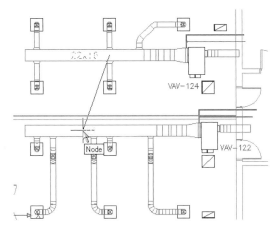

The label appears not to copy. This is because the copy is anchored to the original duct.

    d.  Select the label, right-click, and click Switch Curve, and then select the lower main duct.

    The label displays on the lower duct.

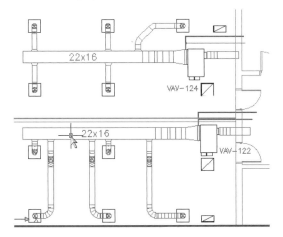

### Add Labels on Another Layer by Setting the Layer Key Override

4.  Change the default layer key style:

    a.   From the Desktop menu, click Layer Management ➤ Layer Key Overrides.

    b.   In the Major Field under Overrides, click the ellipses button.

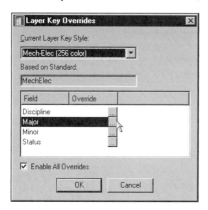

    c.   Scroll down and click Pipe, "Piping" and then click OK twice.

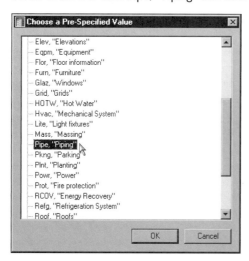

        The override places the labels on the M-Pipe-Label instead of the default M-Anno-Label. You can also set the discipline override to P for Plumbing if you wish.

### Place a Label on the Pipe Using the System Label

5.  Label the hot water pipe:

    a.   Verify that OSNAPs are turned off.

    b.   From the MEP Common menu, click Labels ➤ Add Label.

c. Select the upper red pipe as the curve.

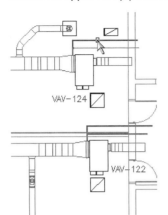

d. The command line prompts you for the style name; enter ? and press F2 to get the AutoCAD text window and you will see the available label styles in this drawing. Press F2 to close the text window.

e. Enter a system label at the command line and press ENTER three times to accept the defaults of manual and 1 node.

f. Then select a point where you wish the label to appear on the pipe.

The hot water system label is placed on the red line. Repeat step 5 to place a system label on the green return pipe. The labels will not hide the 2-line representation of the pipe, but will hide the 1-line pipe representation.

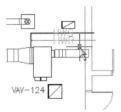

6. Move the label:

a. Select the HWS system label, right-click, and click Offset Node.

b. Click a point above the red line to place the HWS system label.

Remember to turn off the layer override when you are done or everything else that you draw will have this override.

The labels are sized by the drawing and annotation scale set in the drawing setup. The final drawing shows the drawing scale (intended plotting scale) reset from 1/8 to ¼ inches. The labels automatically rescale themselves to be plotted at ¼" = 1'-0" rather than 1/8" = 1'-0".

## EXERCISE 2: USING ARCHITECTURAL DESKTOP ANNOTATION

Architectural Desktop's annotation symbols are accessed through AutoCAD's DesignCenter. Architectural Desktop adds a customization to the DesignCenter interface that allows custom commands to be launched when a piece of 'content' (title marks, detail bubbles, section marks, etc.) is placed in the drawing. All of the annotation that you find under the Documentation ➤ Documentation Content menu launches a command line subroutine. You may view the command by selecting a piece of content, right-clicking, and clicking Edit while in the custom view of DesignCenter. If the symbol is not quite what your office uses, you can simply open up the drawing that stores the content and redefine the blocks in the way that you need them to appear. If you redefine the blocks in your drawing, note that the new content may not overwrite blocks of the same name in your working drawings.

All of the predefined content uses the current text style for the attributes buried in the blocks, so you should make sure your office standard text style is current in your office templates. The predefined content also uses the Drawing Scale and the annotation scale settings in the Drawing Setup dialog box.

**Open:** *ADT annotation.dwg*

### Add a Detail Mark to the Drawing

I.   Place the detail mark:

   a.   From the Documentation menu, click Documentation Content ➤ Detail Marks.

   b.   In the right panel of DesignCenter, left-click on Detail Boundary B and drag it into the drawing.

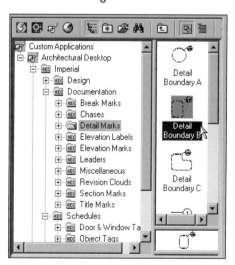

   c. While following the prompts at the command line, pick a lower corner and upper corner for the boundary area.

   d. Make a leader with three pick points: one on the boundary, another to get the bubble into a clear area, and a horizontal point to tie it together, and then press ENTER.

2. Specify the text for the detail mark:

   a. In the Edit Attributes dialog box, enter **M4.5** for the sheet number.

   b. Enter **5** for the Detail Number.

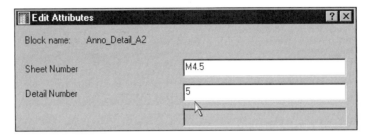

   c. Click OK.

The detail bubble is placed in the drawing.

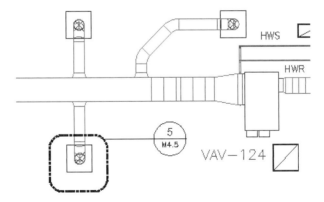

## Add a Title Mark in Paper Space

You can add the title marks in paper space or model space and they will scale themselves appropriately.

3. Place the title mark:

   a. Double-click outside the viewport to get into paper space.

   b. From the Documentation menu, click Documentation Content ➤ Title marks.

   c. Left-drag Title mark A1 into the drawing below the viewport, enter **3** for the number, and then click OK.

4.  Specify the text for the title mark:

   a.  The scale is ¼" = 1'-0" and is read from the Scale tab of the Drawing Setup dialog box.

   b.  Enter **Partial Mechanical Plan** for Title, and then click OK.

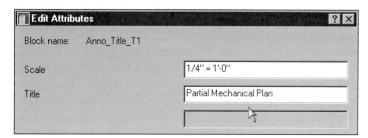

   c.  Pick a point for the end of the line of the title.

   The title mark is placed in the drawing.

## LESSON 3: DISTRIBUTING YOUR MECHANICAL DESIGN

Now that you have created a building model, added schedules and annotation, you will need to communicate the information in the building model to the other members of the design team. Traditionally this is done by sharing the .DWG file, or by printing the drawing set out and distributing the hard copies, or both. This lesson provides important information that will help you share your building model data.

This lesson does not contain any exercises, but instead contains three sections that provide conceptual information about distributing your Building Systems drawings to the rest of the design team.

This lesson contains the following sections:

*   Building Systems Objects and Proxy Graphics
*   Plotting Your Drawing
*   Saving Your Drawing in Other Formats

### SECTION 1: BUILDING SYSTEMS OBJECTS AND PROXY GRAPHICS

Throughout this book a distinction has been made between AutoCAD entities and Building System objects. Building System objects have many qualities and capabilities not found in standard AutoCAD entities. In particular, objects have multiple display representations such as one line, two line, or model. One way of looking at a display representation is a set of instructions held by the object and passed on to your graphic card by the Building Systems software. A standard installation of AutoCAD does not know how to interpret these instructions and views all the Building System objects as proxy objects. A system variable proxy graphic determines what happens when a Building Systems drawing is open in a plain AutoCAD session. The proxygraphics variable is stored in the drawing and has a value of either 0 (which is the default installed value) or 1.

With proxygraphics set to 0, no geometry of the object is stored in the drawing. When this drawing is opened in plain AutoCAD, all the Building System objects are displayed as a bounding box and text.

With proxygraphics set to 1, the geometry of the objects is stored in the drawing. When this drawing is opened in plain AutoCAD, the object can be seen, but the user has no control over the objects. The last current display representation when the drawing was saved becomes the proxy graphics that are seen by the AutoCAD user. For example, prepare a drawing for a consultant and set proxygraphics to 1. You save a drawing in model view looking at the 3D view of the ducts. When this drawing is opened by the consultant using plain AutoCAD, they will see only the model representation for all objects. If you last saved the drawing in the plan 2-line view, the consultant will only see the 2D representations of the objects. Also, the consultant will have no control over any of the objects in the drawing. They cannot move them, or change their layer color or linetype.

While at first this appears ideal, we don't want our consultants messing with our drawings, and there are ways of configuring the display configuration to show all the Building System objects as a gray color as most consultants would use as a background drawing. However, there are times when a consultant or another person in the office will need to modify our drawings without having to buy a seat of Building Systems.

There are two ways for you to enable other members of the building team to use your drawings:

1. You can request that the other members of your team load the appropriate object enabler. Current object enablers are available for download from Autodesk's Point A website.

2. You can explode the Building Systems objects down into basic AutoCAD lines, arcs, and circles.

### Object Enablers

Object Enablers are the Autodesk solution to distributing drawings to members of the team that do not have access to Building Systems. Object Enablers are a free download. Once installed, the Object Enabler gives a standard installation of AutoCAD the ability to understand the information held in the proxy objects. Users can move, copy, or erase the objects, as well as change their layers and colors.

The following information describes the positives and negatives of using object enablers:

Positive: Everybody on the team should be able to utilize the objects and therefore access drawings that contain objects.

Negative: You cannot get an object enabler to have backwards compatibility; therefore, all members of the team must be on the same version of AutoCAD. You cannot give them a drawing prepared in Building Systems Release 3 and have them load an object enabler for AutoCAD Release 14.

### Exploding Objects

The following information describes the positives and negatives of exploding objects:

Positive: No proxy graphic messages, and everything can be changed to ByLayer. Also, everything will be just AutoCAD lines, arcs, circles, and faces. Your consultants do not need another piece of software to install and maintain.

Negative: You give up everything about the objects, including all of the display control and style control. This can be overcome by copying the drawing into another folder before you explode it. See Appendix B for a detailed exercise in how to explode your drawings.

### SECTION 2: PLOTTING YOUR DRAWING

There could be an entire book written about plotting drawings alone. In fact, there is a plotting book available from AutoDesk's e-store for the 2000/2002 versions of AutoCAD. If your office is not already up to speed on plotting in ACAD 2002, you should purchase the book. If your office is already plotting with ACAD 2002, there are a couple things that you should know about plotting with Architectural Desktop and Building Systems.

The default colors assigned to Architectural Desktop and Building Systems object's layers are held in the layer key style. The layer key style can be accessed on the Desktop menu by clicking Layer Management ➤ Layer Key Styles. All of the layer key styles provided with the software correspond to, and are intended to be used with, the two color table styles AIA (256) Scale 48.ctb and AIA (256) Scale 96.ctb. These two color table styles are installed by default in the folder *C:\Program Files\Autodesk Building Systems 3\Plot Styles*. The AIA (256) Scale 48.ctb is for plotting at a 1/4" = 1'-0" scale, and the AIA (256) Scale 96.ctb with lighter lineweight assignments for files to be plotted at 1/8" = 1'-0". If you are setting up the plot styles for your office, these are a good place to start. We have provided a file on the CD-ROM that may be useful as you set up the plot styles for your office.

**06 BuildingSystems R3 LayerKey to CTB2 charts.xls** is a spreadsheet with three worksheets.

The first worksheet contains the layer key style with the colors, and the other layer settings used by each key. The first worksheet also contains the pen weight assigned to each color in the two CTB files.

The second worksheet maps ACI colors to pen widths for both the AIA (256) Scale 48.ctb scale and the AIA (256) Scale 96.ctb.

The third worksheet is a list of the colors that are used by the Layer Key style. This file has been posted to the AutoDesk Customer Files newsgroup, so feel free to use it or share it as you see fit. It is a good starting place for when you need to adjust the layer key style to meet your office needs.

## SECTION 3: SAVING YOUR DRAWING IN OTHER FORMATS

AutoCAD 2002 provides a couple of different ways to save your drawing information in different formats. Most of these are operated through a set of drivers that are installed with the software. These plot drivers are seen by AutoCAD as plotters, but by outputting the file to different formats you can save your drawing as a Drawing Web Format (.DWF) file for email, or for posting the file to a Website or as a Web page. You can also save your drawing as a Joint Picture Expert Group (. JPG) or Portable Network Graphics (.PNG) in order to attach the drawing to an email or insert the drawing into another document.

The Drawing Web format maintains all of the information in vector format. This format is useful when you are creating Web pages, or need to provide drawing information to a third party that you do not want to give your original drawings to. These drawings are much smaller in size than their originals, and this makes them easy to transfer over the Web. Volo View Express is a free download from Autodesk that will view and plot these types of files, as well as provide a Web viewer plug-in when you install it.

You create a .DWF file by plotting to one of the built-in drivers provided for this purpose. On the Plot Device tab of the Plot dialog box, you can choose the plotter configuration .DWF ePlot.pc3 (optimized for plotting) or DWF eView.pc3 (optimized for viewing), and then specify the location and name of the file, and a DWF is then created for you.

Using the plotters to create .PNG files or .JPG files works the same, and is a good alternative if you just need to send somebody a picture of the drawing; however, these files lack the true scaling ability that the vector-based .DWF file has.

Another choice that has become popular in the last several years is to set up a .PDF driver in your Windows printers, and utilize it to create .PDF documents from your files. The PDF can be viewed with Adobe Acrobat® Reader. Adobe Acrobat® Reader is a free download from Adobe. If you choose this option, the .PDF can be plotted, but cannot be modified from its original form.

# KEY CONCEPTS: COMMUNICATING YOUR DESIGN

Objects contain display information that Building Systems reads and then displays on your screen.

You must specify the flow direction using the Set Flow command before the Show Flow command can display the flow arrows.

The default direction of the arrow is dependent on the direction the duct was originally placed in the drawing, but can be changed at any time.

You can create your own custom flow arrows by adding a standard AutoCAD block to the display representation of the duct.

You can add a custom block to any display representation for ducts, pipes, and fittings, but not to MvParts (multi-view parts).

Insulation or lining may be applied to the ducts as they are drawn, or after they are placed in the drawing.

You can change how the insulation is depicted or displayed by changing the display properties of the duct using Duct Properties.

All vertical segments of duct are always assigned the drop symbol by default.

A duct's rise/drop symbol is controlled by a Rise/Drop style accessible in the Style Manager.

When you add vanes to an elbow, the vane gets added to the fitting style and all elbows of that size display the vanes as well.

A label is a style-based object, that when placed in the drawing has an anchored relationship with the object it is assigned to.

Any standard AutoCAD text style may be assigned to a label style.

Architectural Desktop's annotation symbols are accessed through AutoCAD's DesignCenter.

# CHAPTER 7

# Creating Catalog-Based Parts

After you read this chapter, you should understand the following:

- how to use the Catalog Editor to create a fitting based on an existing fitting
- how to create and modify a block-based MvPart
- how to create a parametric-based MvPart

## CREATING AND MANAGING CATALOG-BASED CONTENT

As you have worked through this book, you have used many different pieces of content provided with the software. You have used style-based content—labels, schematic symbols, and lines—to create a schematic diagram. Style-based content is stored in the templates or drawings. You have worked through exercises creating and managing this content with the Style Manager. You have also used catalog-based content—Mvparts, ducts, and fittings—to create the mechanical model. While the term *content* is used within Building Systems to describe all of the parts, pieces, styles, and templates provided with the software, in this chapter it refers to two catalog-based Building Systems objects—fittings and MvParts.

Each of the catalogs that are installed with the software contains many parts. Even so, there will probably come a time when you need a part or fitting that is not in one of the catalogs. Building Systems provides two tools to help create and manage parts: the Catalog Editor and the Content Builder. Both of these tools manipulate the contents of a catalog.

### CATALOG EDITOR

If you just need to modify an aspect of an existing part or would like to copy an existing part and use it as a base for a new part, you can use the Catalog Editor. This tool allows you to copy parts from one catalog to another. It also allows you to manipulate values within the catalog directly, changing behavior or sizes of parts directly.

### CONTENT BUILDER

The Content Builder allows you to create parts and add them to the catalog. You can create either fittings or MvParts with the Content Builder. If you are creating an MvPart, you have the choice of creating a block-based or parametric-based MvPart. Fittings created with the Content Builder will always be parametric. You can also use the Content Builder to add or modify sizes of existing parts in the catalog.

### CATALOGS

The catalog itself is an .XML file with the .APC extension. APC stands for Autodesk Product Catalog. Just as the .XML file for the part keeps track of the other components, the drawing, and the bitmap, the catalog .XML keeps track of all of the parts within it. The product catalogs can be opened with any .XML editor. However, the Catalog Editor provides several useful tools specifically created for the .APC files.

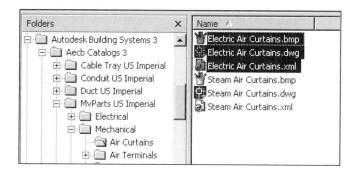

Whether you're working with a fitting or MvPart, there are three components for each part stored in the catalog: the drawing file stores geometric information about the part, the bitmap stores the preview image, and the .XML file holds nongraphical information about the part, as well as the paths to the drawing and bitmap.

## CATALOG PART BITMAP COMPONENT

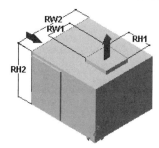

The bitmap is a 200 x 200 pixel image. Many of the bitmaps have additional information added that provide clues to how the part was created. The connector's width and height variables are located with dimension strings, and the intended flow through the part is annotated with blue arrows.

When a part is added to the drawing, the bitmap appears in the Add Dialog box. When the part is modified with either the Part Properties or the Part Modify dialog boxes, it also appears as a reference.

Although neither the connector nor the flow direction annotation is required, as you add or modify a part, knowing the location of the connector's width and height is extremely useful. The majority of the parts have more than one connector. The dimensions in the bitmap help you keep track of them. The blue flow arrows help you align two parts when you are adding them next to each other.

While the bitmap can be generated automatically from the part using the content builder, the annotation is not. You must add this information using paint or some other image editing software.

## CATALOG PART DRAWING COMPONENT

The drawing component of a catalog part holds the geometry needed to create the part. In Building Systems release 3, a parametric modeler was introduced. The parametric modeler can be used to create either fittings or MvParts. The new Content Builder creates COLE-based parametric geometry. Consequently, you will find two different types of parts in the catalog. Older fittings are based on TOM files (a drawing file with 2D geometry), or created with the new parametric content builder. Likewise, when working with MvParts, there are chapters in the catalog that contain the older block-based MvParts as well as the new COLE-based parametric parts.

An example of this is shown in the dampers chapter of the MvParts catalog. The fire damper part was available in release 2 and is block-based. The new simple damper part is an example of a COLE-based parametric part. These two types of MvParts are indistinguishable while you are working with them in the drawing. However, when you are working with the catalog source drawing, they are very different. The fire damper is created from a collection of AutoCAD blocks. The simple damper is a parametric COLE object.

## FITTINGS

The catalog source drawings for fittings may be of two kinds: TOM files or parametric files. You can see the difference between these types of files when you open them.

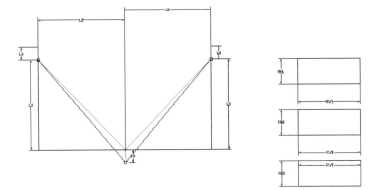

### TOM Fittings

When you open a fitting based on TOM files, the geometry of the part is laid out in 2D lines, held in a block. This example shows a rectangular parts duct fitting. The path of the parts are shown on the left, the shapes of the connectors are held in the geometry on the right. The information about how this geometry is constructed, as well as how the different dimension variables relate to each other, is held in the .XML file of the part.

The construction, layer assignments, and block names are all very specific. We have been told by Autodesk not to modify the geometry of these files, so we pass on this bit of wisdom to you. Do not modify these files. You will probably never need to anyway. Most of the control of these types of fittings is found in the .XML file. The first lesson leads you through several exercises to copy one of these fittings, adjusting the .XML file variables to create a new part.

Another reason not to modify these files is that different fittings may share the same drawing file. Take the example of a tee fitting and a reducing tee fitting. The geometry is the same for both parts. The difference is what happens with the sizes of the connectors in relationship to each other. A regular tee fitting that will have the same size duct on all three connectors is assigned only one connector height (RH1) and one connector width (RW1). This pair of variables is assigned, in the .XML file, to each of the three connectors on the tee. A transitional tee needs to accommodate different size ducts at each of the three connectors. This tee will need additional RH2, RW2, RH3, and RW3 variables in the .XML as well, to assign to the other connector.

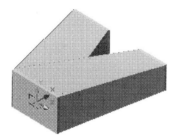

## Parametric Fittings

When you open a parametric fitting, it will appear as a solid object. You can modify the base geometry of these parts, but should only do so with the parametric modeler built into the Content Builder. As with the TOM fitting, the .XML file holds most of the relational variables for the part, and is where you will most often be working to copy this fitting to create another fitting.

### MULTI-VIEW PARTS (MVPARTS)

MvParts in the catalog can also be of two types: block-based or parametric-based. It is important to understand that when they are placed in a working drawing, there is no difference between these two types of parts—it is only in the catalog that these two types of MvParts differs. Both the DWG and the XML files differ for the two types of MvParts.

## Block-Based MvParts

The drawing file stored in the catalog for a block-based MvPart contains all 16 blocks for each size of the part. For example, the fire damper has 10 different sizes, so the catalog's source drawing will have 160 blocks. The .XML file is used to store the block names in relation to the size part. The .XML file also holds the connector sizes and locations, as well as the layer key to be used with the part.

When you use the Add MvPart command to add a 20x20 fire damper to a working drawing, the software uses the blocks from the catalog's source drawing file together with the information from the .XML file to create a new MvPart style for the 20x20 fire damper in the working drawing.

## Parametric-Based MvParts

Unlike the block-based MvPart, a COLE-based parametric part stores the geometry of the part in the catalog's source drawing file as a COLE object, the same way a parametric fitting is stored. The COLE object is a parametric object created from parametric lines, profiles, and extrusions. Although similar to AutoCAD lines, polylines, and ACIS solids, these are specific objects used by the parametric engine to create MvParts and fittings.

When you use the Add MvPart command to add a simple rectangular damper to a working drawing, the software generates the blocks the MvPart style needs in the drawing based on the geometry of the COLE objects in the catalog's source drawing.

### CATALOG PART XML COMPONENT

The .XML file holds instructions about the geometry held by the source drawing files. This includes information such as the layer key, sizes of connectors, part type, and subtype, and much more.

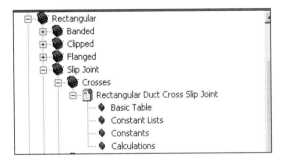

Each .XML file has the same four components: a Basic Table, a Constants List, Constants, and a Calculations area. Each of these part parameter areas contains specific data for the part. Because the base geometry is different for the parametric fitting or MvPart than it is for the TOM fitting or block-based MvPart, the information in each of these areas differs also.

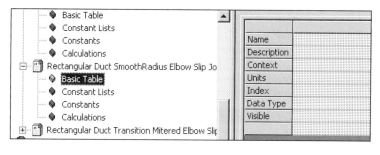

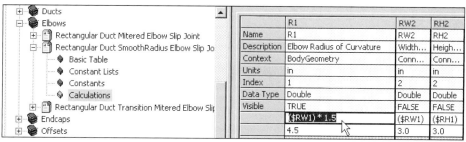

For a fitting, the Basic Table parameters may be empty while the Calculations area holds data for the connectors and relational calculations between the variables of the part. In this example, the value for RW1, the width of connector 1, is multiplied by a factor of 1.5 to derive the radius of the elbow fitting. If you look at a block-based MvParts .XML file, its Basic Table is populated by information tying together all of the blocks that make the part, and the Calculations parameters area is empty because there are no calculated values needed for an MvPart, since everything is derived from the blocks themselves.

These three components—the bitmap, drawing and .XML files—must be accounted for with each piece of content that you create. Both the Catalog Editor and the Content Builder have built-in tools to help you create and manage these files. The lessons in this chapter were designed to give you hands-on experience in creating and managing these files.

## LESSON OBJECTIVES

In the first lesson of this chapter we will use the Catalog Editor to create a new catalog, copy it, and modify a TOM-based fitting. The second lesson will work with block-based MvParts. The chapter will end with a lesson on working with the parametric parts. This chapter assumes you have worked with the software and understand the basic nature of fittings and MvParts, including how they are used, the nature of connectors, and the automatic insertion of fittings. You should be comfortable with the display representations of the objects, as you will be modifying display representations for the parts in this chapter.

This chapter contains three lessons. The lessons are broken down into one or more exercises. Most exercises have a corresponding drawing file that is located on the CD-ROM included with this book. The exercises in this chapter are long, and there are no good stopping places mid-exercise. Please allow a minimum of one-half hour per exercise to permit you to work through the exercise from start to finish.

Lessons in this chapter include:

- **Lesson I: Working with the Catalog Editor**
  In this lesson you will use the Catalog Editor to create a new catalog. You will also use the Catalog Editor to create a new duct fitting based on an existing fitting in another catalog.

- **Lesson 2: Working with Block-Based MvParts**
  In this lesson you will modify an existing block based MvPart—the Perforated Face Neck Ceiling Diffuser - 12 Inch Neck - 48x24 Face—to show a different block in plan view when it is inserted in a drawing. You will also create a new MvPart—a low profile evaporative cooling tower—and add it to the catalog of MvParts.

- **Lesson 3: Working with Parametric COLE-Based MvParts**
  In this lesson you will create a duct silencer using the parametric part builder, adding it to the catalog.

For the exercises in this chapter, you will be working on sample catalogs and parts created for these exercises. Before you start adding or modifying parts on your own, make a backup of the installed catalogs and parts by copying the folder *C:\Program Files\Autodesk Building Systems 3\Aecb Catalogs 3* and all its contents into another location.

## LESSON 1: WORKING WITH THE CATALOG EDITOR

The Autodesk Product Catalog files organize the parts and all of their component pieces. Of the three components—the bitmap, the drawing, and the .XML—the Catalog Editor will give you direct access to the .XML file. The Catalog Editor is a stand-alone application. The executable *AecbCatalogEditor33.exe* is installed in the *C:\Program Files\Autodesk Building Systems 3* folder. You can also launch this application from the MEP Common menu by selecting Content Tools ➤ Catalog Editor.

When you work with the Catalog Editor, you will be working outside of the Building Systems environment. The catalog file is an .XML file with the APC extension. This extension allows the Catalog Editor to find and register it. Although you can work with any of the APC files in any XML editing software, the Catalog Editor is an XML editor designed specifically for use with the Autodesk Product Catalog files, and has unique features you will not find in other XML editors.

This lesson contains four exercises. It is important that you do not work with your production catalog when working through these exercises. The first exercise here will lead you through creating a catalog. You will then copy parts into this "test catalog." When you have completed the exercises in this lesson, you will have copied a radiused elbow fitting and modified it to use a .75 multiplier of the width to derive the radius of the radiused elbow.

This lesson contains the following exercises:

- Viewing Information with the Catalog Editor
- Creating a New Catalog
- Copying Parts from One Catalog to Another
- Copying and Modifying Parts within a Catalog

 **Note:** These exercises build on each other. Please work through them in sequence. Before you work on any of the exercises in this lesson, please copy the folder *07 Exercise Catalogs* and all of its contents from the CD-ROM onto your C: drive, and remove the read-only attribute from this folder and all of its files.

### EXERCISE 1: VIEWING INFORMATION WITH THE CATALOG EDITOR

A catalog is the organizing structure for the XML, DWG, and BMP files for each part. The catalog itself is an XML file with an extension APC, which stands for Autodesk Product Catalog. The Catalog Editor opens the Autodesk Product Catalog file and allows you create, modify, and delete the parts in the catalog. In the first part of this exercise, you will use the Windows Explorer to see how the different catalogs are organized. In the second part of this exercise, you will open some of the existing catalogs to familiarize yourself with the layout of the Catalog Editor. You will then create a new catalog in the next exercise.

#### View the Organization of the Catalogs with Windows Explorer

1. Start Windows Explorer:
   a. Right-click on the Start button of the task bar.
   b. Select Explore.

2. Browse to and view the contents of the folder *C:\Program Files\Autodesk Building Systems 3\Aecb Catalogs 3\MvParts US Imperial\*.
   Notice the catalog here: *All Installed MvParts US Imperial.apc*.

3. Expand the Mechanical folder. Notice the catalog here, *Mechanical MvParts US Imperial.apc*.

244

 **Note:** Each catalog will register the parts that are contained in the subfolders below it. The All Installed catalog reads all of the parts in the folders below it. The Mechanical MvParts catalog reads only those MvParts stored in folders below the *Mechanical* folder where the catalog file is located.

If you create new parts, or substantially modify the catalogs to meet the needs of your office, be sure to archive the parts you modify. When new releases come out all you need to do to get the new parts into the new install is to copy your office parts into a folder below the location of the new installations catalog file. When you regenerate the catalog, the catalog will read your parts and include them in the catalog supplied with the software.

4. Expand the *Dampers* folder and view its contents.

   Here you see the three components for each of the parts—the bitmap, drawing, and .XML file. Each folder in the tree will become a chapter when inserting parts or using the part builder.

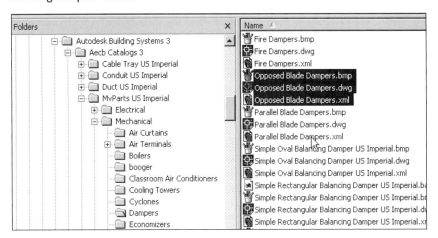

5. Browse to and expand the folder *C:\Program Files\Autodesk Building Systems 3\Aecb Catalogs 3\Duct US Imperial\Rectangular\Slip Joint\Tees*.

   The duct catalog contains both the catalog definitions for the duct as well as the duct fittings used with that shape or type of duct. Unlike the MvParts that store the drawing, XML, and bitmap in the same folder, the duct catalog keeps all the drawings in one folder, all of the bitmaps in another, and the XML files for each part in a different folder. This is because some of the fittings utilize the same geometry, but apply different calculations to get the resultant part. The calculations are stored in the .XML file. An example of this is the Tees. The Bullhead, Straight, and Transition Tees all use the same base geometry, and hence the same catalog reference drawing file, but apply different calculations to create the part in the target drawing.

6. Close Windows Explorer.

**Start the Catalog Editor**

7.  Start the Catalog Editor.

    a.  Launch Building Systems 3 software.

    b.  From the MEP Common menu, click Content Tools ➤ Catalog Editor.

**View the Different Catalogs**

8.  Open the MvParts Mechanical Catalog.

    a.  From the Catalog Editor File menu, click Open.

    b.  Browse to the folder *C:\Program Files\Autodesk Building Systems 3\Aecb Catalogs 3\MvParts US Imperial\Mechanical.*

    c.  Select the *Mechanical MvParts US Imperial.apc.*

        The Mechanical MvParts is a smaller catalog within the larger All Installed MvParts catalog. It contains just the mechanical MvParts.

9.  Open the All Installed MvParts Catalog.

    a.  From the Catalog Editor File menu, click Open.

    b.  Browse to the folder *C:\Program Files\Autodesk Building Systems 3\Aecb Catalogs 3\MvParts US Imperial.*

    c.  Select the *All Installed MvParts US Imperial.apc.*

**View the Catalog Entries for a Block-Based vs. COLE-Based MvPart**

10. View the block-based Fire Damper.

    a.  Expand the catalog in the left window of the Catalog Editor.

    b.  In this tree view of the catalog, expand down through the Mechanical, Dampers, and Fire Dampers branches.

c.   Select Basic Table in the tree view.

The basic table for a block-based MvPart shows the name of the part and the different sizes of the part. For each size, the basic table shows the connectors for the size, along with its size and location. If you have created a block-based MvPart and just need to adjust a connector size, you can do this here, as an option to working through the Content Builder.

d.   Select Constant Lists in the tree view.

The Constant Lists for a block-based MvPart lists the block associated with each view when the MvPart style is created in a new drawing for each size of the part. The recipe for each line in this table is:

View Name, Display Representation, Block Name, and the name of the view assigned to each the block. The view blocks are standard AutoCAD blocks that are stored in the drawing file in the catalog. We will look at each of these in more depth when we create the block-based MvPart later in this chapter.

e.   Select the Constants, then the Calculation items in the tree view.

The Constants and the Calculations table entries for a block-based MvPart are minimal, holding, for instance, naming information for the part. These two categories are used more with fittings and COLE-based MvParts.

11.  View a COLE-based Simple Rectangular Damper.

a.   In the tree view of the catalog, expand Simple Rectangular Balancing Damper.

b.   View the Basic Table, the Constant Lists, Constants, and Calculations for the COLE-based part.

The COLE-based MvPart is parametric. It does not utilize blocks stored in the drawing to create all the needed view blocks. When a COLE part is placed in a drawing, the parametric part will create the needed view blocks in the MvPart Style in the target drawing. Another difference from a block-based part is that the parametric part relies more on calculations and constants. Although the catalog source drawing holds the geometry of the part, the parametric data for this part is displayed here in the catalog.

**Open and View a Duct and Fitting Catalog**

12. Open a duct catalog.

   a.   From the Catalog Editor File menu, click Open.

   b.   Browse to *C:\Program Files\Autodesk Building Systems 3\Aecb Catalogs 3\Duct US Imperial.*

   c.   Select the *Duct US Imperial.apc* and click Open.

13. View the contents of the catalog.

   a.   Expand the tree through Rectangular, Slip Joint, and Elbows.

   b.   Select the Rectangular Duct Smooth Radius Elbow.

   c.   Look at the three entries at this level for the elbow. The first level of the catalog here contains the path name to the drawing file, the bitmap, and the name of the part.

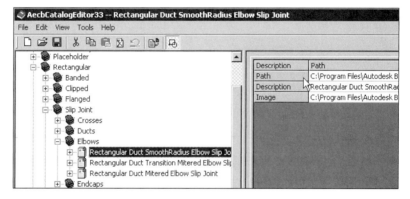

   d.   Look at each of the parameter tables for the elbow; duct fittings have the same parameters as MvParts: Basic Table, Constant Lists, Constants, and Calculations.

The Constant Lists in this case is a list of each size of each variable of the part—RW1 is the width of connector 1, RH1 is the height of connector 1, and A1 is the included angle of the elbow.

The Constants area of the .XML file holds the shape of each connector.

The Calculations parameter of this part is used to assign values to the width and height of the connectors. Here, the connectors are assigned a variable. This fitting will not act as a reducer, so both connectors on the part can use the same size. The Calculations parameters allow you to assign variables to any aspect of the part—in this case the dimensions of the connectors. For this fitting, RW2 is set to be equal to RW1, and the value of RH2 to the value of RH1. With this arrangement, the duct entering the fitting will determine the size of the duct exiting the fitting. The data from the Constant List parameter table is applied to the variables in this chart to create each fitting as it is added to a drawing. Variables are also used in the name of the part. For example, the value FormatNumber ($RW1,1) reads the value of RW1 and places it as a text string in the name of the part. This way, you do not need to create a discrete name for every different size of this part.

The Catalog, although perhaps intimidating at first, is logically organized. Understanding the parameter tables for each of the different types of Building System parts will give you a head start when you create and modify the parts in later exercises in this chapter.

### EXERCISE 2: CREATING A NEW CATALOG

In addition to viewing the contents of a catalog, you can use the Catalog Editor to create new catalogs. When you are creating new parts, it is a good idea to work in a separate catalog. If you develop your parts in a separate catalog, the regeneration and testing time is shorter, and it keeps partially-done content out of the working catalog and out of the production drawings. In this exercise you will create a catalog that you will use for the remainder of this lesson. Just as each part has an image preview, each catalog too can be assigned a preview image. We have provided a .JPG file for you to use with the catalog you create in this exercise. The subfolder *C:\07 Exercise Catalogs\07 New Catalog* where you will create the new catalog contains a .JPG file that you will assign to the catalog at the end of this exercise

Before you start this exercise, you should have copied the *07 Exercise Catalogs* folder onto your hard drive.

**Start the Catalog Editor**

1.  Start the Catalog Editor:

    a.  Start a session of Building Systems.

    b.  From the MEP Common menu, click Content Tools ➤ Catalog Editor.

**Create a Duct Catalog**

2.  Create a new catalog with the Catalog Editor:

    a.  From the Catalog Editor's File menu, click New.

    b.  Select Duct_Component for the Domain Name.

    c.  Enter **Duct Exercises** for the catalog name.

    d.  Enter **Examples** for the description.

e.   Click the ellipse button; browse to and select the folder *C:\07 Exercise Catalogs\07 New Catalog* and click OK to return to the Catalog Editor.

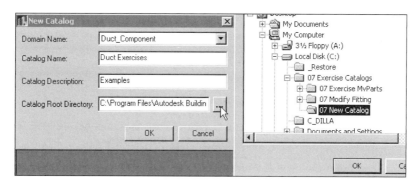

3.   Assign the provided bitmap for the catalog by typing **ExerciseDuctCatalog.JPG** at the end of the image path name that defaults to the catalog path location.

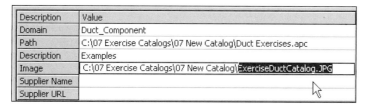

The catalog is created and opened. The preview image for the Catalog appears in the preview box. The image file is not crucial for a catalog. However, if you do not assign an image file to the catalog, you will get a non-crucial error message each time you open the catalog file telling you that it does not have one. The catalog will still work without a preview image.

In the next exercise, you will populate the catalog by copying parts from an existing catalog. If you are going on with the next exercise, you can keep this new catalog open.

### EXERCISE 3: COPYING INFORMATION BETWEEN CATALOGS

Once you have created a catalog, you can populate the catalog with parts by copying parts from another catalog. In this exercise, you will have two sessions of the Catalog Editor open. You will use the copy and paste functions of the Catalog Editor to copy the rectangular duct parts from the installed catalog to populate the Duct Exercise catalog you created in the last exercise. If you are continuing from the last exercise, skip down to step #2.

### Open the Duct Exercise Catalog and the Installed Duct Catalog

1. Open the Exercise Catalog:

   a. Start a session of Building Systems.

   b. From the MEP Common menu, click Content Tools ➤ Catalog Editor.

   c. From the Catalog Editor's File menu, click Open.

   d. Browse to *C:\07 Exercise Catalogs\07 New Catalog*.

   e. Select the *Duct Exercises.apc*. and click Open.

2. Start a second session of the Catalog Editor and open the installed duct catalog:

   a. From the MEP Common menu, click Content Tools ➤ Catalog Editor.

   b. From the new session's Catalog Editor's File menu, click Open.

   c. Browse to *C:\Program Files\Autodesk Building Systems 3\Aecb Catalogs 3\Duct US Imperial*.

   d. Select the *Duct US Imperial.apc*.

 **Note:** You can position both sessions of the Catalog Editor on your screen by selecting one session on the windows task bar, holding the control key, selecting the second session, then right-clicking and selecting Tile Windows Horizontally.

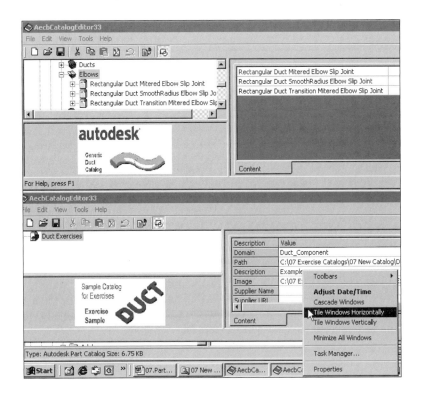

**Copy and Paste Part of the Installed Catalog into the New Catalog**

3.   Copy the Rectangular Duct from the installed catalog:

   a.   In the left pane of the installed Duct US Imperial catalog, expand the tree through Rectangular, Slip Joint, and select Elbows.

   b.   Click the Copy icon in the toolbar.

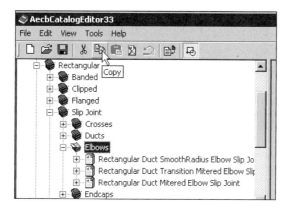

4.   Paste the rectangular parts into the new catalog:

   a.   Move to the Catalog Editor session with the Duct Exercise catalog open.

   b.   Pick the Paste icon on the toolbar.

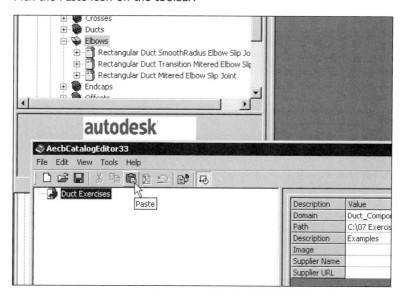

Be patient with this command—it is not instantaneous. This is not a simple copy and paste as would happen with Windows Explorer. When you copy and paste with the Catalog Editor, it must perform several operations on each part. First, the Catalog Editor duplicates the .BMP, .DWG, and .XML files for each part in the branch you are copying. Next, the catalog generates a unique Global Unique Identifier (or GUID) for each of the new parts. Then the Catalog Editor rewrites the path name within the .XML file for each of the new parts updating the path names to the drawing files and bitmaps for each part.

c. Click "Yes to All" when the file overwrite message occurs.

When you copy parts or branches from one catalog to another, there will be a couple of messages about overwriting files. One for the DWG files, and one for the BMP files. In the first lesson, we pointed out that more than one fitting might use the same reference drawing. When you copy a catalog branch like this, each part will look for all of the pieces it needs to create itself—in this case, some of the drawings and bitmaps are duplicated. This is fine.

5. Close both sessions of the Catalog Editor, saving changes to only the Duct Exercises catalog.

If you view the contents of the *C:\07 Exercise Catalogs\07 New Catalog* folder, you will see that it contains a folder named *BMP* with three bitmaps, a folder named *DWG* containing two drawings, and an *Elbows* folder containing the three .XML files for the three parts that were in the elbows chapter you copied from the installed catalog.

Using the Catalog Editor to copy parts is not just a simple copy function. You could copy the parts folders and files to a subfolder below the exercise catalog using Windows Explorer. When you regenerate the catalog, the parts would show up. However, when you copy and paste with Windows Explorer, the GUIDs will be the same as the original part. Additionally, the path names in the .XML files will be incorrect. This will cause problems when you start using this part in your working drawings. The process you just went through, copying and pasting with the Catalog Editor, will create unique GUIDs, and update the paths in the parts .XML file.

Not only can you copy parts between catalogs, you can duplicate parts within the same catalog to generate similar parts. In the next exercise, you will copy and modify a fitting in the Duct Exercise catalog we just created.

### EXERCISE 4: COPYING AND MODIFYING FITTINGS WITH THE CATALOG EDITOR

We have on the newsgroup an occasional request to have a rectangular smooth radiused elbow with a smaller radius than the default with a radius of 1.5 times the width of the duct. In this exercise, you will copy the smooth radius elbow and modify the calculations with the Catalog Editor creating a new part—a smooth radiused elbow with a radius of .75x the width.

Before you start this exercise, you should have copied the *07 Exercise Catalogs* folder to your C: drive. This folder contains the *Duct Exercise 04 catalog* that we will use with this exercise. As you develop your parts, you will probably find it advantageous to work with a smaller catalog for several reasons. Developing parts outside of the larger working catalog will prohibit you from corrupting the working catalogs. The smaller catalog will regenerate and test much faster than a catalog with many parts.

## Open the Duct Exercise Catalog

1.  Open the Exercise Catalog:
    a.  Start a session of Building Systems.
    b.  From the MEP Common menu, click Content Tools ➤ Catalog Editor.
    c.  From the Catalog Editor's File menu, click Open.
    d.  Browse to *C:\07 Exercise Catalogs\07 Modify Fitting*.
    e.  Select the *07 Modify Fitting.apc.* and click Open.

## Copy and Rename the Fitting

2.  Copy the fitting:
    a.  Expand the Elbows.
    b.  Click on the Rectangular Duct Smooth Radius Elbow Slip Joint.
    c.  Click the Copy icon on the toolbar.

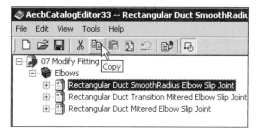

3.  Paste the fitting:
    a.  Click on Elbows in the tree view of the Catalog Editor.
    b.  Click the Paste icon on the toolbar.

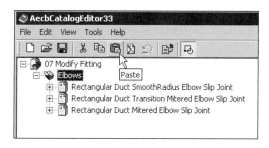

4. Rename the fitting:

   a. Right-click on the new fitting and select Rename.

   b. Enter **SmoothRadius .75 width Elbow** in the name field (the / mark is not allowed here).

5. Copy and paste this value into the description field on the right side of the Catalog Editor also.

6. Expand the fitting to see the fittings components by clicking the plus sign next to the part name.

## Modify the Fitting Values and Test the Catalog

7. Modify the fittings radius calculation:

   a. In the right window of the Catalog Editor, place the cursor over the edge of the columns and drag to expand their values.

   b. Double-click on the next-to-last value in the R1 column to open it for editing.

   c. Change the value from ($RW1) * 1.5 to **($RW1) * .75**.

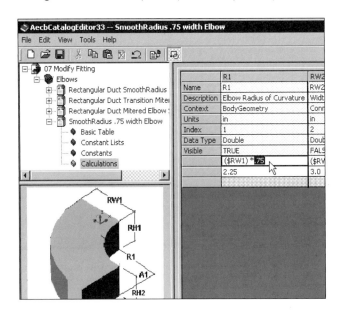

8. Click the Save toolbar icon to save the part and catalog changes.

9. Test the catalog with the Tools menu selection Test Catalog.

   You can now use either the Catalog Editor or Windows Explorer to move the new .75 radiused elbow part into the working catalog folder. If you use the Catalog Editor to copy this part back into the folder for the rectangular duct slip joint elbows, the Catalog Editor will adjust the drawing and bitmap paths for you. If you cut and paste this part in the folder for the rectangular duct slip joint elbows using Windows Explorer, you will have to open the *Duct US Imperial.apc* and manually update the bitmap and drawing locations with the Catalog Editor.

## LESSON 2: MODIFYING AND CREATING MVPARTS

Much has been described about MvParts in other areas of this book. To create and modify MvParts, you should already understand what an MvPart is, and how it behaves in a drawing. In addition, you should understand the difference between MvPart Styles as they occur in the drawing and the source drawing in the catalog.

### MvParts Display Representation

To understand the display aspects of MvParts, think of a nested block with the ability to be selective about which of the internal blocks is displayed in the drawing. MvParts are comprised of several standard AutoCAD blocks. Which block is displayed in the drawing is determined by the display system used by Building Systems. When you view an MvPart from the top, a 2D-plan block of the part is displayed. When you view the MvPart in an isometric view, the 3D-model block of the part is displayed. In addition to the view direction-dependent display of the MvPart there are five different display representations for the MvPart: 1-Line, 2-Line, Model, Plan, and schematic displays. Each of these display representations can have a different block for each of the view directions. So the part can display a different block in the 1-line top view than the 2-line top view. In general, an MvPart will have 16 blocks that make up all of the display representations and view directions. The 1-Line and 2-Line representations have blocks for the view directions: top, bottom, right, left, front, back, and 3D, creating a total of 14 blocks. The Model representation uses the 3D block for all view directions. The Schematic representation uses the schematic block for all view directions for a total of 16 blocks all together. The Plan representation is not used by the content installed with the software.

### MvParts Styles

Block-based MvParts store these display blocks in the catalog source drawing file. Typically, all the different sizes for a part are stored in the same source drawing. The fire damper has 10 different sizes, with 16 blocks for each size, and so the source drawing has 160 blocks. When you add the part to a working drawing, these blocks are pulled from the source drawing and put together to create the MvPart style in the working drawing. The .XML file guides which blocks are assigned to each display representation's view direction. The blocks, although named in the source drawing file, become anonymous blocks in the working drawing.

Unlike the Schematic Symbols, the source drawing does not contain an MvPart style. The style is created when the part is added to the drawing. If you want to modify how a part looks in a particular drawing, you can modify the parts style in that drawing. If you want to modify how the part appears in all the drawings you create, you will need to modify the catalog source drawing.

If you modify the catalog source drawing, the parts will not affect an existing drawing that already has that part. For example, let's say you open the source drawing for a return air grille and redefine all the plan blocks to show a slash through the part. If you add a grille to a drawing where this part already exists, the old part will be used. There are two reasons for this. The grilles MvPart style exists in the drawing and will not be overwritten. The AutoCAD block definitions that are assigned to the grille's MvPart style also exist and will not be overwritten. In order for you to add the redefined part you must rename the MvPart style in the working drawing.

## MvParts Types

MvParts are categorized by type and subtype. The type of an MvPart is a fixed list in the software. The subtype can be created as you create a part. An example of this are the Air Terminals. Both the ceiling diffuser and the return air grill have the type Air_Terminal, but they have different subtypes. Types and subtypes are very useful when used in conjunction with Quick select when creating a selection set of MvParts.

This lesson contains two exercises, both of which focus on working with MvParts using the Content Builder. In the first exercise you will modify an existing MvPart, creating a new view block of a ceiling diffuser as a rectangle with an X through it to replace the standard plan view block used by the software. In the second exercise, you will create an MvPart from scratch using the Content Builder. In both of these exercises, you will be working with a sample catalog created for these exercises.

This lesson contains the following exercises:

- Modifying a Block-Based MvPart
- Creating a New Block-Based MvPart

### EXERCISE 1: MODIFYING A BLOCK-BASED MVPART

Block-based MvParts store all the needed view blocks in the catalog's source drawing. If you want to change how the part appears in a working drawing, you can simply edit the MvParts style in that working drawing. If you want to change how the MvPart is created in all new drawings you will need to change the catalog's source drawing file. The exercise "Changing the Display of an MvPart" in Chapter 6 showed you how to assign a different block to the MvPart style in a working drawing.

In this exercise, you will edit the catalog source drawing for a ceiling diffuser, adding a block to be used as the plan block. After the source drawing has the revised plan block you will use Content Builder to reassign the block assignments for the views. Because the Content Builder will only look at the current catalog, you will need to make the Exercise MvPart catalog current before using the Content Builder. The illustration below shows the before and after images of the part in top view.

BEFORE        AFTER

You will be modifying the *07 Exercise MvParts catalog* and its source drawing file. The catalog and source drawing are contained in the *07 Exercise Catalogs* folder that should already be on your C: drive.

**Create a New Plan Block for the Diffuser**

1.  Start a session of Building Systems.

2.  Open the source drawing in the *07 Exercise Catalogs* catalog:

    a.  From the Building Systems File menu, select open.

    b.  Browse to *C:\07 Exercise Catalogs\07 Exercise MvParts\Diffusers*.

    c.  Select the *Exercise Ceiling Diffusers.dwg*.

3.  Create a block for the new plan representation:

    a.  Verify the Layer is set to 0 and change the color and linetype to **ByBlock** and **ByBlock**.

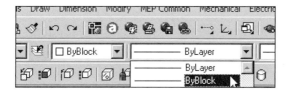

    b.  Use the Rectangle command to create a rectangle from 0,0 to 24,24 and then draw lines attaching each corner in a cross.

c. Use the Draw menu to select Block ➤ Make and then select all of the objects.

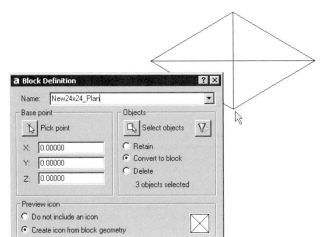

d. Use the base point of 0,0, name the block **New24x24_Plan** and then click OK.

4. Save and close this drawing.

## Set the Current MvParts Catalog to the Exercise Catalog

5. Reset the MvParts Catalog:

a. From the Tools menu, select Options.

b. On the Building System Catalog tab, click the browse button next to the MvParts path.

c. Browse to *C:\07 Exercise Catalogs\07 Exercise MvParts* and select the *07 Exercise Mvparts.apc*.

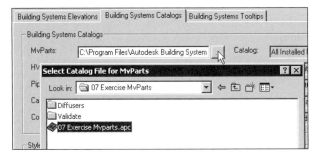

d. Click Open and verify that the description of the MvPart catalog reads, "MvParts for Exercises" and click OK.

### Start the Content Builder

6.    Start the Content Builder:

a.    From the MEP Common menu, click Content Tools ➤ Content Builder and verify that the Part Domain is MvParts.

b.    Expand the Diffusers in the tree view and click on Exercise Ceiling Diffusers as the part to modify.

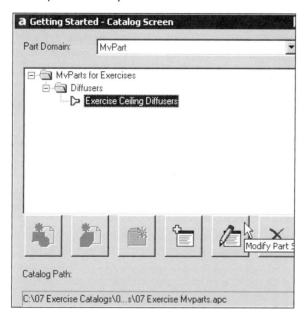

c.    Click the Modify Part size Icon to bring up the wizard.

### Work through the Wizard to Modify the Part

7.    Reassign the New 24x24_Plan block to the top view of the 24x24 diffusers:

a.    Skip the Behavior tab where you can adjust the Layering, Type, and Part behaviors and click the Blocks and Names tab.

b.    Click the Perforated Face Neck Ceiling Diffuser- 8 Inch Neck - 24x24 Face in order to highlight that row.

c.    Scroll to the column Top Block, click on the name to activate a drop-down menu that accesses all the blocks in this source drawing and select the block New 24x24_Plan.

Repeat this for the two other neck sizes that utilize the 24x24 plan size.

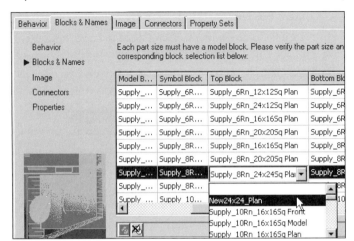

d.  Click the Generate Blocks button.

You can verify that the blocks are assigned in the Views dialog box. Use the Part Size name to select one of the 24 x 24 sizes and then click on the numbers 1 and 8 to see the assigned blocks for One_Line_Top and Two_Line_Top, respectively.

e.  Click OK twice.

8.  If you are done with the exercises for now, remember to reset the MvParts catalog back to the working catalog—either in *C:\Program Files\Autodesk Building Systems 3\Aecb Catalogs 3\MvParts US Imperial* or the mechanical subfolder of this location. If you are going to continue with the next exercise, keep the catalog set at the exercise catalog.

You have now modified this diffuser to use the block you created in the beginning of the exercise. Note that the catalog definition for this part has been modified. The catalog drives the creation of the part's MvPart Style as it is placed in a drawing. If you have previously added this part to an existing drawing, the style will already exist. In that case, adding this modified part will not change any existing style in the drawing, the additional parts placed in an existing drawing will continue to use the existing style definition for the diffusers. However, in any new drawing that this part is placed, the definition will be read from the catalog, and the new plan block will be used with the part.

### EXERCISE 2: CREATING AN MVPART

In this exercise, you will create an MvPart for a low profile evaporative fluid cooler from scratch. The cooler is a piece of equipment based on a Baltimore Aircoil cooling tower. This piece of equipment comes in seven different models with differing capacities. The seven models are configured in one of three housings. All of the models have the same footprint, differing only in height. Because of the nature of the housing, the entire series can be modeled in four blocks. Three ACIS solid blocks represent the three different housing sizes, and one 2D schematic block represents the footprint of any of the models. In the drawing you will open there has been some work done already. We have already created the ACIS solid blocks and the schematic block for you.

In this exercise, you will open the drawing with the blocks in it and use the Content Builder to create the MvPart. The Content Builder has a wizard-like interface that will take you through the steps creating the part.

This exercise will build a new part and place it in the current catalog. If you are continuing from the previous exercise, your MvParts catalog should be set to the Exercise MvParts catalog. If you have set the catalog to one of the standard installed catalogs, please reset this to the Exercise MvParts catalog before proceeding.

 **Open:** *07 Evap Cooler MvPart.dwg.*

### Create a New Chapter for the Part

1. Start the Content Builder:

   a. From the MEP Common menu, click Content Tools ➤ Content Builder.

   b. Verify that the Part Domain is MvParts and Click on the MvParts for Exercises line in the Content Builder window.

   c. Click the New Chapter icon (the folder), enter **Evaporative Fluid Cooler** for the chapter name and then click OK.

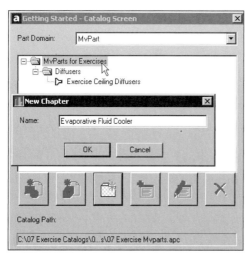

Once you have created the new chapter, the New Part icons become available.

### Create a New Part and Specify its Behaviors

2.  Create a new block-based part in this catalog:

    a.  Click the New Block Part icon (on the left).

    b.  Enter **Low Profile Liquid Cooler** for the name.

    c.  Click on the description line, and the description will fill in for you based on the name.

    You can change this if you like. The name is used as the style name when this part is added to a drawing. The description is shown when you add a part. You can schedule either of these values using tags. One of the label styles created for you in the template will read this name into the label. Keep these facts in mind when you name your parts.

    d.  Click OK to access the Content Builder's block wizard interface.

3.  Set the behavior of the part in the Behavior screen of the wizard:

    a.  Use the Type drop-down menu to select Cooling Tower.

    b.  Verify that Cooling Tower automatically fills in for the layer key and then enter **Evaporative** for the subtype.

    c.  Leave the Breaks Into and Anchors check boxes unchecked and click Next.

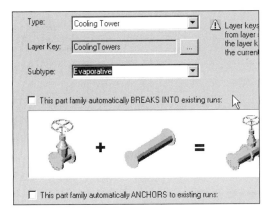

### Establish the Part Sizes and Generate Blocks from the ACIS Model Block

4.  Generate the blocks for the MvPart:

    a.  On the Blocks and Names page, click the Add Part Size icon (the small icon below the Part Size Name column) and select *4d_LowProfileChiller_Small* for the Model block.

    b.  In the symbol block line, select *4d_LowProfileChiller_Schematic*.

    Repeat, adding *4d_LowProfileChiller_Med* and *4d_LowProfileChiller_Large* as the model blocks for new sizes, with the *4d_LowProfileChiller_Schematic block* for the Symbol Block.

You can also double-click inside the Name column to change the names of the parts you are creating.

 **Note:** Add all of the sizes you have created before you go on to the next page of the wizard. You are working in a drawing outside of the catalog, but when you are done with this process, the Content Builder will create a source drawing file for the piece of content you are building. This source drawing will have only those blocks listed on this page. If you create just one part size, you will have to add the blocks for the other sizes manually to the source drawing as we did in the first exercise.

At this point, the model block and schematic blocks are assigned to each of the part sizes. The red block names do not exist, and will be generated for you based on the ACIS solid model blocks in the drawing.

   c.   Click the Generate Blocks button.

5.   Review the block assignments.

   a.   When the blocks are generated, it will bring up the Views window for you to look at each of the parts you are generating. You can use this page to see that there is a view block assigned to each of the views. If you are creating more than one size part, review each of the parts by changing the part with the Part Size drop-down at the top of this dialog box. The blocks are generated with a HIDE command. If you want the part to have a different plan symbol that would be generated by the HIDE command, you should create it before you launch the Content Builder. You can then assign it to the top view for each representation here in the Views dialog box.

 **Note:** Be careful of randomly picking on the views page. Although it is not obvious you are doing so, picking any of the different Display Representations at the lower-left portion of this dialog box will assign the view block shown to the representation you highlight, instead of the representation it should be assigned to.

   b.   Click OK in the Views window.

   c.   Click the Next button on the Blocks and Names window.

6.   Generate an image:

   a.   Click the radio button "Generate an image based on a model block from the SW isometric view."

   b.   Select *4d_LowProfileChiller_Large* for the model block to generate the view from.

   c.   Click the Generate button.

If you are creating a part that has a critical flow direction and would like to show flow arrows in the preview image you can edit this image later, or you can create your own image before you start this process and browse to the image from this page.

   d.   Click the Next button.

### Add Connectors to the Chiller

7. Add an intake pipe connector to the chiller:

   a. On the connectors page, select the line Low Profile Liquid Cooler.

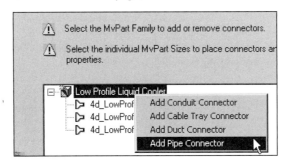

   b. Right-click and select Add Pipe Connector.

   c. In the Part Family Connector Properties page, type **Intake** for the name.

   d. Change the flow direction to **In** and then click OK.

8. Repeat the actions in step 7 to add the outflow pipe connector to the chiller. Name the connector **Outflow** and set the flow direction to **Out**.

   Two connectors have been added to each of the sizes of the part. Because different sizes of the part will have the same connectors, you add connectors to the MvPart family and they are added to all of the part sizes. These connectors then need to be located and sized for each part in the family.

### Locate and Size the Connectors

9. Establish the position of the intake connector on the small size chiller:

   a. Right-click on Connector 1 under the *4d_LowProfileChiller_Small* to access the Edit Connector page.

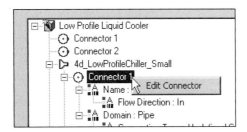

   b. The connector editor is opened.

c.  Click on the ellipses next to the position to pick a point in the drawing.

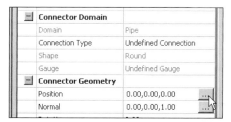

d.  Use the Center OSNAP to select the center of the top cylinder extending from the left side of the model.

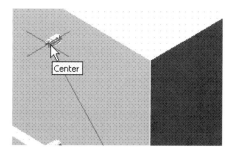

10. Set the Normal variable and size for the part:

a.  Type **–1,0,0** for the Normal variable.

You can select this on screen as well, but for this example, it is easier to type it in.

The Normal variable is a directional vector for the connector. The Normal is set *not* to the flow direction of the pipe, but to the direction in relation to the part that the pipe will be drawn away from the part you are creating—in this case, in the negative X direction. It is used when you start a pipe on this connector to give the pipe segment a direction in relationship to the part.

b.  Type **3** for both the Diameter and Nominal Diameter for this part.

c.  Click OK to return to the MvPart Builder

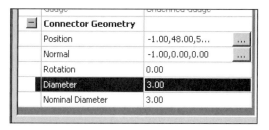

11. Repeat, adding location, size, and Normal data for the small size chillers connector 2. Use the outside center of the lower-left cylinder for the position variable. Enter **−1,0,0** for the Normal and **3** for both the Diameter and the Nominal Diameter.

    If you care to have this part at the end of the exercise to use in your office, you can continue adding the connector information to the medium and large size parts. Use a 4-inch diameter connector for the medium size part, and a 6-inch diameter connector for the large part. If you are not going to use this part, you can just continue with the exercise without adding information for the medium and small part sizes. The larger size parts will just use the default values for the connectors.

**Adjust the Properties of the MvPart**

12. Click the Next button.

13. Adjust the Properties:

    a. You can use this function if you have made any errors, or just want to check the values you have entered on the previous four pages of the wizard. Select a part to view the information that will be written to its .XML file. The Properties pages are broken down into three categories: Calculations, Parameter Configuration, and Values. To edit the values of a block-based part, use the Values tab; most of the editable information will be here. What cannot be edited here are the block assignments and views—you have to go through this process again to do this, or use the back key before you finish the part.

 **Note:** You can add schedule data by adding columns to the Parameter Configuration. Schedule data is added with the following information format—the column name must equal the property set definition name, and the value must equal the property name. When this part is tagged with a tag that uses the same property set, the schedule data should be read from these values on the part.

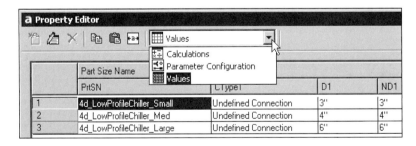

    b. Select the *4d_LowProfileChiller_Small* and select the edit properties button in the lower-left of the dialog box.

    c. Click Finish.

Congratulations, you have created a low profile liquid evaporative cooler MvPart! This part has now been added to the current catalog. You can now access this part and add it to new drawings just as you would any other MvPart in the catalog.

## LESSON 3: CREATING PARAMETRIC PARTS

Working in the parametric mode is a bit different from working in AutoCAD. In the parametric modeler, there are no menus. All the functions are generated with a right mouse click. The base environment of the parametric modeler uses work planes to establish the orientation of the geometry you create. Everything you draw will be created on a work plane. You will select a work plane and right-click to create geometry. This will not be standard AutoCAD geometry, but rather COLE objects. It is important that you use COLE geometry and not standard AutoCAD geometry while working in the parametric interface. Once you have the outline for a part created with COLE lines and profiles, you will extrude the profiles with modifiers to give the part mass. The profiles and extrusion modifiers are all linked back to the work plane where they originated. The next step is to add connectors and refine the connector's relationship to the extrusion modifier with the use of COLE dimensions. Dimensions in the parametric modeler become the basis of the variables used for the part, and will control part size, connector size, and any calculated relationship between different elements of the parametric part.

### Planes

Work Planes are the base of all geometry you create with the parametric modeler. There are 10 different planes you can add to the model to establish and control the geometry.

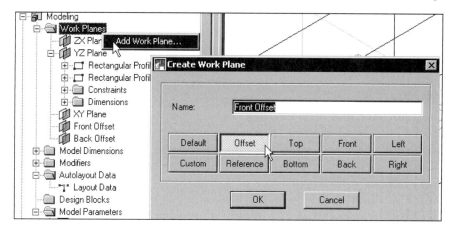

- Default establishes three work planes for you to work on: an XY, a YZ, and an XZ plane.

- Custom allows you to establish the plane by entering in coordinates for the X, Y, and Z location to establish the plane. You can create planes at angles to the world UCS with this option.

- Offset planes reference an existing plane and accept a distance from the original plane to establish their location. Offset planes are useful because they establish a parameter variable you can use to control sizes of 3D geometry linked to the plane.

- Reference planes use the face of a 3D modifier component on an existing plane to create a new plane. This is useful if you want to create a plane on any of the surfaces of a 3D modifier.

- Top, Front, Left, Bottom, Back, and Right planes are just what their names indicate.

## Geometry

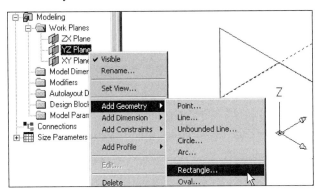

After the preliminary planes are established, you will add geometry. Reinforcing the idea that all geometry is plane controlled, you must select a plane and right-click to add any geometry. You can add basic geometry such as lines, arcs, circles, or derived geometry such as projected geometry that will project edges of modifiers onto a plane. When you are working with geometry, you cannot move the geometry once it is added to the work plane. You use the parametric variables to modify the relationship of a piece of geometry to another geometry. You must give up thinking about distances in terms of inches between lines. Remember that you are creating a parametric part. The distances between different geometries will change as the part size changes. You can use AutoCAD lines, arcs, and circles to help you place the COLE geometry, but these layout lines should be erased before the part is put into production. There is an exception to this rule. The Design Block is used to assign a schematic block to the part. The Design Block can be standard AutoCAD entities tied together in a block.

## Profiles

Profiles, like geometry, are linked to a specific work plane. Profiles are created the same way COLE geometry is, by right-clicking on a work plane in the tree view of the parametric modeler. The choices of profiles are circular, rectangular, oval, or custom. The custom profiles allow you to create a shape using the COLE lines and arcs, and then select this geometry to create a profile. The 3D COLE modifiers are all based on profiles.

## Constraints

Constraints establish relationships between the COLE objects. When you add constraints, you will select one piece of geometry to constrain another. As the original geometry changes, the constrained geometry will also change, based on the constraints applied. Be very careful with constraints. It is very easy to overly constrain the geometry, and less easy to back out of the assigned constraints. While you can see the constraints by selecting them in the tree, as you design it is not easy to keep track of the geometry to which you have already assigned constraints. Additionally, connectors and dimensions have their own internal constraints. You may start a part with several constraints and get no error messages, but when you add a dimension or connector later find that the part is over constrained. You can create parts without constraints, and for your first few parts, you may find it easier not to work with constraints at all. For these parts, you will use the parametric variables of the dimensions and connectors to create the relationships between the components of the part. On the positive side, constraints allow you to create parts with fewer variables. Fewer variables means less to keep track of at the end of the part design when you are establishing the relationships between the parametric variables of the part.

## Modifiers

Modifiers create the mass of the MvPart. The three different types of modifiers—extrusion, path, and transition—all use a COLE profile as their base shape. The extrusion offers several different methods of extruding the profile shape. The following exercise uses the Mid-plane to extrude a profile to either side of the base work plane. The From–To extrusion allows you to specify other work planes to use as the start and end planes for the profile shape. The Path modifier extrudes a profile along any of the linear COLE elements. The Transition modifier uses a different start profile than end profile.

You may use Boolean adding and subtraction on collections of modifiers to obtain complex shapes. Be careful with using the Boolean functions on multiple shapes that will have coplanar faces. When you add connectors to the part, it is difficult to identify which modifier the connector is selecting. Try to plan the part so that you avoid coplanar faces. In the duct silencer you will make in the exercise in this lesson, our first attempt used a Boolean subtraction of the modifiers to create a shell. We found it easier to create solid shapes to represent the silencer. Although the part will be solid, we could not think of a reason why we would ever want to see a section through the part, and accepted that it would be a solid part. This allowed us to shift the end planes of the modifiers away from each other, making it clear to which modifier the connector was added.

## Dimensions and Model Dimensions

Dimensions within the parametric editor are not like standard AutoCAD dimensions. Adding a COLE dimension will add a parametric variable to the part. This allows you to establish relationships between the different COLE geometries that make up the part. For example, when you create a duct connector on the part, it will add a variable for the width RW1 and a variable for the height RH1. You have the ability to set these connector variables equal to any of the COLE dimensions in your part. This fundamental relationship allows the parts to be parametric in nature. You will create a list of sizes for the variables RH1 and RW1. You can then assign RW1 or RH1 to any of the dimensions you establish for the part. As the part size changes, the size is passed through the connector variable to the dimension and then to the COLE geometry changing the physical size of the part. Dimensions will always be created in the same work plane as the geometry it is referencing.

Often you will need a dimension for a modifier that is perpendicular to the work plane. The model dimension takes care of this situation. Added with a right-click from the Model Dimension folder, you can apply this dimension type outside of the work plane that created the 3D extrusion, path, or transition modifier.

## Model Parameters

Model Parameters are created as you add dimensions, model dimensions, offset work planes, and connectors to the drawing. Any parameter can be a variable, such as the connector sizes RH1 and RW1 that are assigned values held in the XML. Parameters can also be set equal to other parameters in the part. In the example you will work through you will assign the length of the part to be equal to the width of the duct. Any type of relationship between the components of the part is done through the model parameters.

## Size Parameters, Tables, Lists, and Copying from Excel

When the part is created, you will also create a table of sizes to apply to the part. Within the Edit Part Size Dialog box there are three different pages: the Parameter Configuration page, the Calculations page, and the Values page. Each of these pages contains different controls for the parameters. You can also add parameters in the Edit Part Sizes dialog box. Each parameter can be a Calculation, a Constant, a List, or a Table.

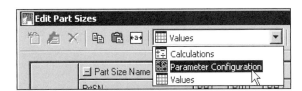

An example of a calculation-type parameter is the radius of the rectangular smooth radius elbow fitting that was modified in the exercise, "Copying and Modifying Fittings with the Catalog Editor," in lesson 1 of this chapter. The calculation RW1*1.5 was used to determine the value of the parameter R1 that became the radius of the elbow.

The Constant type of parameter is used for a dimension or value that does not change. In the exercise, you will create a constant parameter "Name," to hold the constant text string "Duct Silencer" used in the full name of the part.

List type parameters contain a list of variables that can be assigned to the parameter. The List type of parameter is commonly used for the fittings in Building Systems. For each connector's parameter on a fitting, the .XML file contains a list of values that can be applied to the width, and another list that can be applied to the height. The List type of parameter allows all possible combinations of the list variables that will be presented when the part is added. A fitting with the values 10", 14", and 18" for the width and 20", 26", and 30" can be added in any combination of the width and height. When this fitting is added, you will have a choice of nine sizes of fittings.

 **Tip:** If you are creating a fitting type of part, you can open one of the existing fittings and copy the list of sizes to the clipboard and paste the list into your part. You do not need to enter every possible value manually. You may also copy and paste from an Excel spreadsheet into the list parameters.

Table type parameters, like the List parameters, allow you give multiple values to the same parameter. The difference is that the table parameter will not allow cross combinations. A part with the values 10", 14", and 18" for the width and 20", 26", and 30" will present three size choices when added. The exercise uses the table parameter to create three sizes of the duct silencer.

This lesson contains the following exercise:

- Creating a Duct Silencer

### CREATING A DUCT SILENCER

In this exercise, you will create a duct silencer using the Content Builder's parametric part builder. The parametric part builder may be used to create fittings, or as in this case, MvParts. This is a long exercise, roughly broken up into the following steps:

- create a new chapter in the MvPart catalog, and give the part a name and description
- give the part a type, layer key, and subtype
- create work planes
- create profiles
- add constraints and/or dimensions
- create modifiers by extruding the profiles
- add connectors
- add any other relational dimensions needed
- adjust the existing model parameters, using the calculator as needed
- generate a preview graphic
- set the part options
- validate and save the part

As in AutoCAD, there are many ways to get to the end result. The steps in this exercise are written as much to demonstrate different features of working with parametric parts as to get to the final product. Working in the parametric modeler takes getting used to, and there are many features not touched on in this exercise. However, after creating a few parts, the modeler will become familiar enough to allow you to create new parts quickly.

This exercise will start from scratch, so there isn't a drawing file associated with it. We will be using the *07 Exercise Mvparts.apc* catalog found in *C:\07 Exercise Catalogs\07 Exercise MvParts*. If you are continuing from the previous lesson, then the current catalog should already be set to this catalog. If you have been working with one of the installed standard MvParts catalogs, please set the exercise MvParts catalog as the current one using the Building Systems Catalog tab of the Options page.

**Start the Content Builder and Create a New Chapter**

1.  In a session of Building Systems, start the Content Builder:

    a.  From the MEP Common menu, click Content Tools ➤ Content Builder and verify that the Part Domain is MvPart.

    b.  Click on the MvParts for Exercises line in the Content Builder window.

    c.  Click the New Chapter icon (the folder), enter **Duct Silencers** for the chapter name and then click OK.

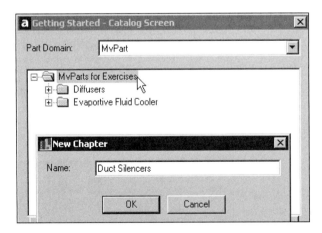

**Start the New Part**

2.  Click the New Parametric Part icon.

    a.  Enter **Rectangular Duct Silencer** for the name.

    b.  Click on the description line to use the name as the description as well.

    c.  Click OK to open the parametric modeler.

**Set the Part Configuration**

The Part Configuration sets the basic properties of the part. The four values here are the part Name, Class, Type, and Subtype. You will see the Name value when you add the part, and it is open for editing here. The Class value of the part is static, set from the choice you made when you started the Content Builder. The Type variable is a drop-down list of preset types. This list is hard-coded and cannot be changed. The Subtype allows user entry. Both the type and subtype are useful for creating selection sets from all the MvParts in the drawing using Quick Select.

3. Set the Type and Subtype:

   a. In the tree view, expand the part configuration.

   b. Use the drop-down list to select Filter from the list.

   c. Double-click the bottom entry of the Part Configuration—Undefined—to open this subtype variable for editing. Type in **Silencer** here and press ENTER to register this value in the line.

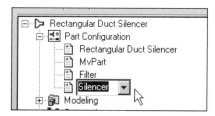

**Create the Work Planes**

The first step in drawing a parametric part will always be to create a work plane on which to draw the COLE geometry. Here you will create the default set of three work planes.

4. Establish the default work planes:

   a. In the tree, expand the *Modeling* folder.

   b. Right-click on the Work Planes and select Add Work Plane.

   c. Click Default and then click OK.

5. From the View menu, select 3D Views ➤ SW Isometric.

6. Turn off the work planes you are not using:

   a. Expand the *Work Planes* folder.

   b. Right-click on the ZX plane and click Visible to uncheck it and turn this plane off.

   c. Repeat for the XY plane.

## Add Profiles

The Profiles you add here will be extruded later to form the body of the part. You could draw geometry such as COLE lines, arcs, and circles, and convert this to profiles later. For a simple part, just start with the profiles directly as it is done here. When the profiles are in place, you will establish some constraints and dimensions to establish relationships between the profiles.

7.  Set the view to the work plane by right-clicking on the YZ plane and click Set View.

8.  Add profiles to the YZ plane:

    a.  In the tree view, right-click the YZ Plane, and select Add Profile ➤ Rectangular.

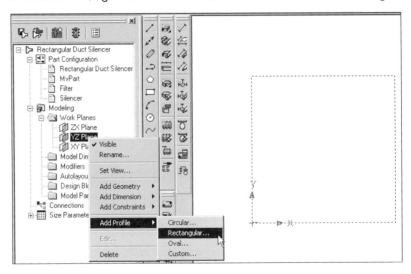

    b.  Type **0,0** at the command line for the first point.

    c.  Type **12,12** at the command line for the second point.

    d.  Repeat, adding another rectangular profile from **2,1** to **8,8**.

## Add Constraints or Dimensions

Constraints establish relationships between the geometry. In this part of the exercise you will establish constraints making the distances between the two profiles equal. You will start this series by turning off the work plane so you can select the edge of the profile that is coincident with the edge of the plane.

9.  In the tree view, right-click the YZ Plane, and click Visible to turn the work plane off.

10. Add Equal Distance Constraints:

    a.  In the tree view, right-click the YZ Plane, and select Add Constraints ➤ Equal Distance.

The command line prompts you for the COLE geometry to establish the constraints.

b.   Select the left side of the outer profile at the "Select pair 1 geometry 1" prompt.

c.   Select the left side of the inner profile at the "Select pair 1 geometry 2" prompt.

d.   Select the top of the outer profile at the "Select pair 2 geometry 1" prompt.

e.   Select the top of the inner profile at the "Select pair 2 geometry 2" prompt.

11.  Repeat step 10 two more times, constraining the left side pairs to the right side pairs and the left side pairs to the bottom pairs. As you add the constraints, the geometry of the profiles changes to meet the constraints imposed on them.

**Adding Dimensions**

At this point in the part creation, you may add dimensions. Dimensions also are used to establish relationships between the geometry of the part, and act as numeric constraints. When you are working with constraints and dimensions, you may sometimes generate a part that does not behave in the way you expect. If you create a part and the parameters do not work the way you expect, try re-creating the part, but leave out the dimensions or some of the constraints. For the part we are creating, it would appear that we could now apply a dimension to the left vertical pairs of lines and have control of the distance of all of the offsets. At the time of this writing, adding the dimension at this point over constrains the part and causes problems with the connectors you will add later. Because of this, we will add a few dimensions after the connectors are attached.

**Add Two Offset Work Planes**

As you added the profiles, they were added to the YZ plane. The 2D COLE geometry and COLE profiles are attached to the work plane they are created in. Work planes can also be used to establish the start and endpoints of the 3D COLE Modifiers. In the next few steps, you will add a couple of Offset Work Planes. The Offset Work Plane is a special type of work plane that has a parameter variable associated with it. You will use the work planes as the start and endpoint for one of the extrusions. Later in this exercise, you will use the parameters that control the offset work plane to control the length of one extrusion modifier.

12.  From the View menu, select 3D Views ➤ SE Isometric.

13.  Make the XY Plane visible.

14.  Add offset work planes in front of the YZ plane:

a.   In the tree view, right-click the *Work Planes* folder, and select Add Work Plane.

b.   Click Offset, name this plane **Front Offset** and then click OK.

c.   Click on the YZ plane in the drawing as the reference work plane.

d.   Drag the cursor to the right and place the plane in the drawing. Distance is not important at this time, as we will modify it later.

15. Repeat this process, adding a second offset work plane named **Back Offset** to the left of the YZ Plane.

16. Expand the *Model Parameters* folder. Each offset plane you add adds a parameter you can use to control the geometry of the part.

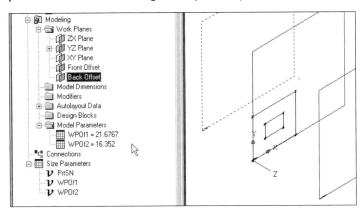

## Create Modifiers from the Profiles

The next step is to create modifiers from the profiles. The modifiers will be the 3D body of the MvPart. In this section of the exercise, you will use two different methods to create the solid body of the part. Both profiles will be extruded using the extrusion modifier. The inner profile will be extruded using the Offset work planes as the start and endpoint to create a core to which the duct will attach. The outer profile will use a mid-plane extrusion to create a shell. The shell and core could both be created using the mid-plane extrusion, or even a path type of modifier. In this exercise, we lead you through creating both types of modifiers just to introduce you to the differences between them.

17. Create a Mid-Plane extrusion modifier from the outer profile.

    a. Right-click on the *Modifiers* folder and select Add Extrusion.

    b. In the drawing, click on the outer rectangular profile and choose Mid-Plane for the type.

    c. Enter **10** for the distance and then click OK.

18. Create a Plane-to-Plane extrusion modifier from the inner profile.

    a. Right-click on the *Modifiers* folder.

    b. Click on Add Extrusion.

    c. In the drawing, click on the inner rectangular profile and choose From-To for the type.

    d. Use the drop-downs to select From = BackOffset, To = FrontOffset and then click OK.

19. Rename the modifiers:

   a.   Expand the *Modifiers* folder and click on each of the extrusions. As you click on the extrusion, it will highlight in the drawing window.

   b.   Right-click on the longer extrusion and click Rename.

   c.   Type **Core** for the name.

   d.   Repeat on the shorter extrusion, naming it **Shell**.

   The modifiers do not add any parameters to your part. You will see that you have parametric control over the Plane to Plane extrusion with the variables WPOf1 and WPOf2 that control the offset planes. You will have to add a model dimension to the core modifier to have a variable you can use for the parametric control over this section of the silencer.

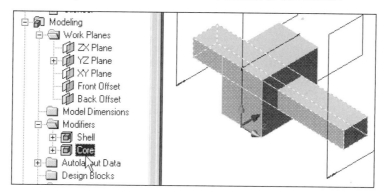

## Add Connectors to the Modifier

Connectors can only be added to a Modifier that exists in the drawing. The connector is the active component of the part when it is placed in a working drawing, determining the duct size that is connected to it. In this next part of the exercise, you will add a connector to the core modifier that extends between the two offset work planes. When you add connectors, the connector will bring with it new parametric variables. A rectangular connector will add RW1 and RH1. When you add the second connector to the same modifier, RW2 and RH2 will be added, and their values automatically set to equal RW1 and RH1. When you add a connector you will be asked for an end of a modifier to place the connector on, and then a height and width for the connector. You will select the lines of the profile that created the core extrusion to establish the dimensions of the connector. This establishes a constraint, setting the width and height of the profile (and hence the core extrusion) to the RW1 and RH1 parameters. This is where you can get in trouble. If you have overly dimensioned or constrained the geometry you associate with the connector, the connector parameters may not be able to control the modifier. Before you start this next set of steps, make sure your view is set to 2D wireframe so you can select the lines of the profile that create the core modifier.

20. Add a connector to a modifier:

    a.  Right-click on the Connections item in the tree and select Add Connection.

    b.  Move the cursor in the drawing to see the connector stick at each end of the extruded modifiers.

    c.  Select the far right position to place the connector on the right face of the Core modifier. Watch the command line—the connector needs a number, a height, and a width.

    d.  Press ENTER to use the default 1 as the connector number.

    e.  Set the height of the connector by selecting the top and bottom line of the inner profile and then place the dimension within the inner profile.

    f.  Set the width of the connector by selecting the vertical lines of the inner profile.

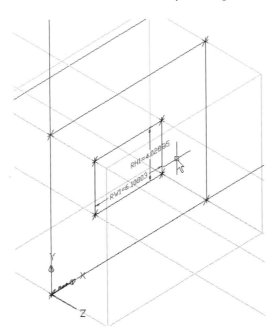

21. Repeat step 20, adding a connector 2 to the other end of the core modifier.

When you add a connector to the second end of a modifier, you are not prompted for a connector width or height; it just uses the RW1 and RH1 from the first connector. You can see this if you look at the list of Model Parameters. There RH2 is set to equal RH1 and RW2 to equal RW1.

22. Set the connector type:

    a.  Expand the Connections in the tree view, if not already visible.

    b.  Right-click on the *Connections* folder and select Edit Connections.

c.   Change both connection types to Duct.

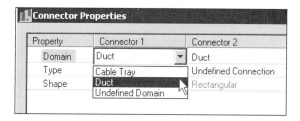

### Add Additional Control Dimensions

The next step is to add any other control dimensions to the drawing. In this case, you are going to add a dimension to the profiles. This dimension will control the reveal between the core and the shell. Because you constrained all the edges, you just need to add one dimension to control all of the edges. You will then add Model Dimension to the shell. Currently you have parametric control over the length of the core with the Offset Plane parametric variables WPOf1 and WPOf2. The model dimension will give you parametric control over the length of the shell. In general, duct silencers are approximately square, so we are going to assume the length of the shell will be equal to the entering duct width. Once the model dimension has been added, you will set this dimension equal to the RW1 dimension using the Model Parameters.

23.  Isolate the profiles:

a.   In the tree view, right-click the YZ Plane, and click the Visible to turn it off.

b.   Repeat for any other visible planes.

c.   Repeat for the two modifiers, turning them off as well.

24.  Add the profile offset dimension:

a.   In the tree view, right-click the YZ Plane, and select Add Dimension ➤ Horizontal Distance.

b.   Select the left side of the outer profile at the "Select first geometry" prompt.

c.   Select the left side of the inner profile at the "Select second geometry" prompt.

d.   Pick a point outside the middle of the left side to position the dimension.

e.   Enter **1** for the dimension. The dimension is added, to the left side. The Equal Constraints make all of the other edges 1" as well.

The Distance option works the same way as the Horizontal Distance, but just reads the value from the drawing, and does not prompt you for a value. There is no difference in the type of parameter added.

25.  Make the two extrusions and the two offset planes visible.

26. Add a Model Dimension:

    a.   Right-click on the *Model Dimensions* folder and select Add Distance.

    b.   Select the Shell modifier (the shorter of the two).

    c.   Pick a point to the left of the model to place the dimension.

    d.   The Model Dimension is added to the Model Parameters as LenB1.

27. View the dimension variables by expanding the *Model Parameters* folder. Each distance that you added to the drawing creates a model parameter in the drawing. Each of the different types of parameters will have unique variable names. The Horizontal Distance command created variable LenA1. Dimension commands will create a LenA# variable. The Model Dimension created variable LenB1. The Offset work planes will create WPOf# variables, and connectors will create variables with the R prefix.

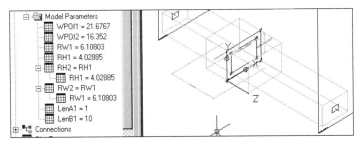

## Set the Parametric Relationships

As you have added dimensions, offset work planes, and model dimensions, you have been creating parametric variables. You will now use the Equation Assistant tool to establish the relationships between these parametric variables. For clarity, you will modify the parameters in the order they were created. First, you will set the length of the core, determined by the parameters for the offset planes WPOf1 and WPOf2. The core width and height are controlled by the connector parameters RW1 and RH1. You will set these values to constants to establish a starting value for the part. They will be assigned more values later when you set the size parameters. The offset between the core and the shell is adjusted with the LenA1 variable. The shell width and height are taken care of with the constraints you placed on them together with the dimension LenA1. The length of the shell is the model dimension LenB1. You will assign LenB1 to be equal to the connector width R1.

28. Right-click on the *Model Parameters* folder and click Edit.

29. Adjust the core modifier length by setting WPOf1 and WPOf2 to equal half the value of RW1 + 2.

    a.   In the Model Parameters dialog box, click WPOf1 and then click the Calculator button.

    b.   Click the left parenthesis button and then click the Variable button.

    c. Click the variable RW1 and use the calculator buttons to set **(RW1*.5)+1** for the value.

    d. Click OK to return to the Model Parameters dialog box.

30. Repeat for WP0f2 using the same value **(RW1*.5)+1**, or set WPOf2=WPOf1 to do the same thing.

31. Use the calculator to set RW1 to **20** and RH2 to **20** as starting values for the part.

32. Set the distance between the inner and outer profiles by adjusting LenA1.

    a. Click on LenA1 in the Model Parameters window.

    b. Click the Calculator button.

    c. Enter **1/2** for the value and then click OK to return to the Model Parameters dialog box.

33. Use the calculator to set the shell length by changing Model dimension LenB1 equal to RW1.

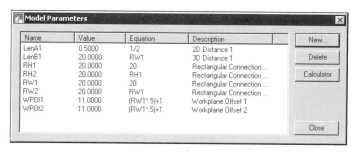

As you change the parameters of the part, the model in the drawing updates to fit the new relationships.

34. Click Close in the Model Parameters dialog box to return to the parametric modeler.

35. Click the Save Part family icon to save your work to this point.

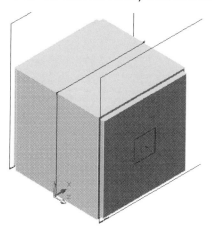

**Adjust the Size Parameters**

You have many parameters in the part already. The parameters you adjusted in the Model Parameters dialog box are all Calculation type parameters. These parameters establish the relationships between the components of the parts. The next to last step is to adjust the size table of the part. In the Size Parameters, you will add more sizes to the part and create a name for the part that will read the size into the part name.

36. Change RW1 and RH1 to Table-type parameters.

    a.   Right-click on the Size Parameters in the tree and click Edit Size Parameters.

    b.   In the Edit Part Sizes, verify the page is set to Parameter Configuration.

    c.   Change the RH1 and RH2 from Constant to Table to allow more than one dimension for these values.

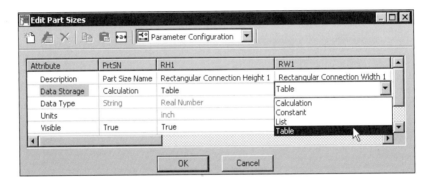

37. Add values to the tables for the width and height.

    a.   Change the page of the Edit Part Sizes dialog box to Values.

    b.   Click the 20" value under RH1 then Click the Add Icon.

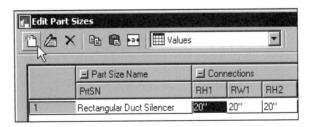

    c.   In the new size 2, enter **40"** for RH1 and **60"** for RW1.

    d.   Repeat, adding a size 3 that uses **60"** for RH1 and **80"** for RW1.

38. Create a new parameter for the name for the part.

    a.   In the Edit Part Sizes dialog box, change the page to Parameter Configuration.

b. Click the New icon

c. Enter **Name** for the new parameter.

d. Click OK to return to the Edit Part Sizes dialog box.

39. Assign the part name variables to read the part size name.

a. Change the page to the Calculations page.

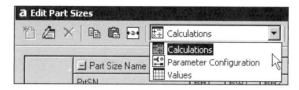

b. Type **Silencer** in the custom column under Name.

c. Enter **FormatNumber($RWI,0) + " X "+ FormatNumber($RHI,0)+ " $Name"** for the calculation of PrtSN. This is a VB format to pull the values of the RWI and RHI into the name.

d. Change the page back to Values. The resulting name should be 10 X 10 Silencer.

## Set the Options

The options hold the layer key, and controls how the part will break into a duct, or anchor to a duct. In this dialog box is a hide part option that allows you to keep it invisible from the Add MvParts dialog box until you are ready to use it.

40. Set the options.

a. Click the Options icon.

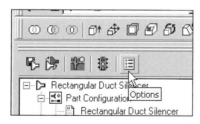

b. Verify that Hide part is unchecked. You can use this setting if you want to save a part in a catalog, but are not ready for it to be used. The part will be hidden in the catalog and will not be able to be seen when adding parts until this is unchecked. Parts that are not valid will have this checked by default. When the part is validated, you are prompted if you want to remove this flag.

c. Select Break into Part to allow the silencer to be placed in existing ducts similar to a damper.

d. Select the ellipses in the layer key and select Filters for the layer key.

41. Set the insertion point.

    a. Expand Autolayout Data.

    b. Right-click on the Layout Data.

       If Autolayout Data will not expand, save the part, close the Content Builder, and open the part again with Content Builder. This seems to force the layout data to become accessible for selection.

    c. Click on Select Placement point.

    d. Use an Endpoint OSNAP to pick the lower front corner of the part.

42. Create a bitmap for the part:

    a. Pick the Generate Bitmap icon on the toolbar.

    b. Pick the SE Isometric view, and then click OK.

43. Validate the Part by clicking on the Validate icon (the stoplight). If the part has been correctly made, the light will be green at the lower-right side of the Content Builder tree. If there are problems with this part, a list of issues will appear on screen.

Congratulations! You made a parametric MvPart! A couple more hints before we close the chapter. We did not create a Design Block for this part. The Design Block is a standard AutoCAD block that is used to represent the schematic symbol of the MvPart. If you test this MvPart, be sure you are not in the schematic layout.

## KEY CONCEPTS: CREATING PARTS AND WORKING WITH THE CATALOGS

MvParts, ducts, and pipes are each stored in their own catalog.

The Duct Catalog contains the fittings as well as the ducts.

For every part in the catalog there are three files stored in the catalog: a bitmap preview image, a source drawing file, and an .XML data file.

You can copy or save as existing catalogs to create new ones to modify.

You should not copy catalog parts with Windows Explorer because the unique ID for the part will not be modified, creating two parts with the same ID. Use the Catalog Editor or the Content Builder to copy parts.

Content Builder allows you to modify or create parametric or block-based MvParts.

Before you create a block-based part you should have a model block and schematic block to represent each different size of the part you will need.

With Parametric parts, you can create multiple sizes of the part based on Model Parameters established as you create dimensions in the parametric drawing.

You can develop parametric parts over time. By keeping the option "hide part" checked, the part will not be visible in the catalog of parts, and so will be prevented from being added to working drawings.

# Calculating Floor Areas

After you complete the exercises in this chapter, you should be able to:

- calculate square footage using spaces
- calculate areas using spaces
- create an area evaluation
- add areas and area groups

## SPACES AND AREAS

There are two AEC objects that will give you the square footage of a room, or part of a building: Space objects and Area objects. Space objects may be present in drawings given to you by the architect as they are used in Architectural Desktop for room and finish schedules. Spaces have a built-in query function that enables you to see a tally of each individual space's square footage, or a square footage total for each style of space in the drawing. Areas and Area Groups are different types of objects that enable you to write out an "Area Evaluation" to an Excel® spreadsheet.

## WHAT ARE SPACES?

Spaces are an Architectural Desktop feature that enables you to calculate the floor area of a building. Calculating the floor area of a building is important in order to estimate the initial per-square-foot costs, as well as to determine preliminary air volume calculations. A space object is comprised of a floor and a ceiling. Both the floor and ceiling portion of a space object contain a thickness value. Spaces also include a height that is calculated by the distance from the top of the floor to the bottom of the ceiling. Spaces can also contain a value for the height above the ceiling.

You can specify the default height above the ceiling space on the AEC DwgDefaults tab in the Options dialog box. You do not need to use this option unless you want to create schedules for volumetric takeoffs.

## UNDERSTANDING SPACES

In Architectural Desktop, spaces serve as a conceptual design tool allowing the architect to lay out areas before walls. Often the space requirements of the building determine the shape of the structure, or, as in the case of a tenant improvement to a shell building, the spaces are more important than the original form of the building.

Spaces have a built-in square footage calculator which you will be using in lesson 1. Although you can specify spaces to schedule any of the above values, this appendix focuses on teaching you how to use the built-in square footage calculator.

Spaces are style-based objects. The Architectural Desktop templates contain the following space styles:

- Standard
- Mechanical room

- Electrical room
- Restroom (men and women)
- Closet
- Conference (large, medium, and small)
- Office (large, medium, and small)
- Workstation (large, medium, and small)

The space styles contain properties that enable you to define the default values for the space. These are accessed through Architectural Desktop's Space Style Properties dialog box.

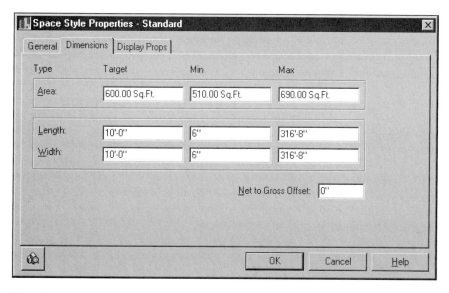

The Dimensions tab enables you to define the constraints of the space, which will be used as a default when adding a space to your drawing. You are not able to exceed the values that you specify on this dialog box when adding a space. However, if modifications need to be made later, you can grip-edit the size of the space. This tab also contains an option called "net to gross offset". The net to gross offset is used when you convert spaces to space boundaries, or start to convert a space plan into a building. You do not need to use this setting to complete the lessons in this appendix.

You use the Display Props tab to specify different hatches for space styles so that you can tell them apart in the drawing. While an architect may want to have many different space object styles, you will create three space object styles to use in this exercise: *Zone 1*, *Zone 2*, and *Zone 3*.

If the architect you are working with is using Architectural Desktop, you may get a drawing with spaces already in it.

**LESSON OBJECTIVES**

This appendix contains two lessons. The lessons are broken down into one or more exercises. Most exercises have a corresponding drawing file that is located on the CD-ROM included with this book.

Lessons in this appendix include:

- **Lesson 1: Calculating Square Footage using Spaces**
  You will open an architectural drawing that contains pre-drawn spaces and use a query feature to see the space sizes. You will also create spaces styles to represent different zones within a building and then add them to your drawing.

- **Lesson 2: Working with Area and Area Groups**
  You will work with a drawing that contains area and area groups in order to familiarize yourself with the area evaluation interface. You will also learn how to write out the square footage to an Excel spreadsheet. You then learn how to add areas and area groups to a drawing.

## LESSON 1: CALCULATING SQUARE FOOTAGE USING SPACES

The following exercise is representative of a drawing file that could be supplied to you by an architect who uses Architectural Desktop's spaces. You should open the drawing file and use the Space Information dialog box to view the dimensions of the space. In the second exercise you will add spaces to represent the zones of the building. You will then grip-stretch the spaces to fill areas. You will also learn how to convert a polyline into a space.

This lesson contains the following exercises:

- Creating and Querying Spaces and Space Styles
- Generating Spaces

**Note:** In Building Systems two Architectural Desktop pulldown menus were removed. You still have access to the functionality of these commands by entering the command at the command line. If you want to add spaces to your drawing, you can type in SPACE at the command line, and are given the command line choices of [Add/Convert/Modify/Properties/STyles/Join/Divide/SWap/Query/Interference]. You could follow through this exercise in Building Systems with command line prompts, but we suggest instead that you open Architectural Desktop before you open these next two exercise drawings so that you have access to the Concept pulldown menu selections.

## EXERCISE I: CREATING AND QUERYING SPACES AND SPACE STYLES

In this exercise, first you will open an architectural drawing that contains spaces in order to view square footages of the spaces in the drawing. You will then use Style Manager to create space styles to represent different zones in the building and add the spaces to your drawing.

 **Open:** *Apdx A Querying Space.dwg*

1. Generate a space inquiry:

    a. From the Concept Menu, click Spaces ➤ Space Inquiry.
    You can also enter **space** at the command line, and then enter **Q** for Query.

    b. In the Space Information dialog box, click the Space Info Total tab.

    The Space Info Total Tab groups and sorts all of the spaces contained in the drawing by style. You can use this tab to view the total square footage of each space style using the Area column.

    c. Click the Space Information tab to view the information for each space separately.

Click to export this information to an access data base file

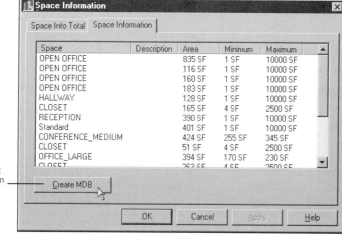

 **Note:** This drawing already contains spaces that were arranged by an architect who provided the base drawing for you to work on. The spaces were placed in the drawing by individual rooms. This exercise shows how you can set up spaces to work in zones rather than individual rooms. In the following steps, you create three space styles – Zone 1, Zone 2, and Zone 3.

2. Create space styles in Style Manager:

    a. From the Concept menu, click Spaces ➤ Space Styles.
    You can also enter **space** at the command line and then enter **ST** for style.

b. In the left panel, select Space Styles, right-click, and click New.

c. In the right panel, double-click New Style to open the Space Style Properties dialog box for the new style.

3. Edit the space style in the Space Style Properties dialog box:

a. Click the General tab and enter **Zone 1** for Name and **Western Perimeter** for Description.

You do not need to change the properties on the Dimensions tab. The space will default to a 10' x 10' area and you will grip-stretch it later to modify the size.

b. Click the Display Props tab and select Space Style under Property Source.

c. Click Attach Override and then click Edit Display Props to open the Entity Properties dialog box.

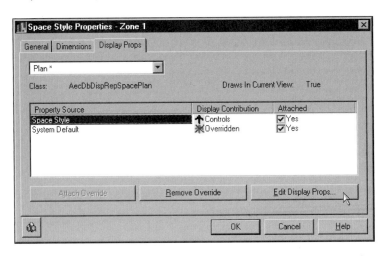

4. Specify a hatch color and pattern for the space:

   a. Click the Layer/Color/Linetype tab, select the word Hatch, and change the color to red.

   You change the color by clicking ByBlock in the Color column and choosing a color or entering a number in the Select Color dialog box and then clicking OK.

   b. Click the Hatching tab, then under Angle change 45 to **315**, and then click OK two times to return to Style Manager.

   You change the hatch angle so that the space appears different than the spaces that are already in the drawing.

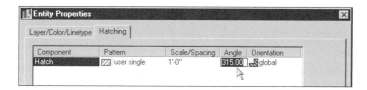

5. Create a second space style:

   a. In the left panel of Style Manager, select Space Styles, right-click, and click New.

   b. In the right panel, double-click New Style to open the Space Style Properties dialog box.

   c. Click the General tab and enter **Zone 2** for Name and **Northern Perimeter** for Description.

   d. Click the Display Props tab, select Space Style, click Attach Override, and then click Edit Display Props.

6. Specify a hatch color and pattern for the second space:

   a. Click the Layer/Color/Linetype tab, select the word Hatch, and change the color to 30 (orange).

   b. Click the Hatching tab, then under Angle enter **315**, and then click OK three times to return to Style Manager.

7. Create a third space style:

   a. In the left panel of Style Manager, select Space Styles, right-click, and click New.

   b. In the right panel, double-click New Style to open the Space Style Properties dialog box.

   c. Click the General tab and enter **Zone 3** for Name and **Core Areas** for Description.

   d. Click the Display Props tab, select Space Style, click Attach Override, and then click Edit Display Props.

294

8. Specify a hatch color and pattern for the third space:

   a. Click the Layer/Color/Linetype tab, select the word Hatch, and change the color to blue.

   b. Click the Hatching tab, then under Angle enter **315**, and then click OK two times to return to Style Manager.

   c. Scroll down in the left panel of Style Manager to view the new space styles: Zone 1, Zone 2, and Zone 3, and then click OK.

 **Note:** You can also copy and paste styles in Style Manager using the shortcut menu.

In the following steps, you add the spaces to the drawing using the styles that you created.

9. Add Zone 1 to the drawing:

   a. In the drawing, zoom into office number 404 (located at the lower left).

   b. From the Concepts menu, click Spaces ➤ Add Space.
You can also enter **space** at the command line, and enter **A** for Add.

   c. Select Zone 1 for style, pick a point at the lower left of office 404, and press ENTER twice to accept the space's default insertion properties.

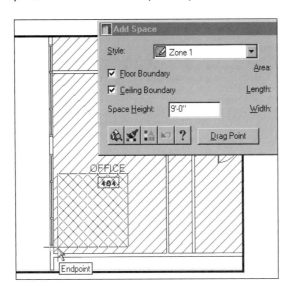

10. Change the size of the Zone 1 space:

    a.  Zoom out to see all three of the offices that are located on the left-hand side of the drawing.

    b.  Select the space and grip-stretch it to fit over offices 404, 405, and 406.

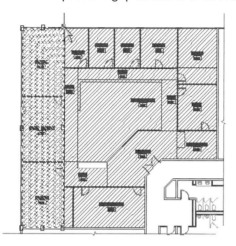

11. Add Zone 2 to the drawing:

    a.  In the drawing, zoom into office number 407 (located at the upper right).

    b.  From the Concepts menu, click Spaces ➤ Add Space.

    c.  Select Zone 2 for style, pick a point at the lower left of office 407, and press ENTER twice to accept the space's default insertion properties.

12. Change the size of the Zone 2 space:

    a.  Zoom out to see all five of the offices that are located across the top of the drawing.

    b.  Select the space and grip-stretch it to fit over offices 407, 408, 409, 410, and 411.

13. Convert a polyline to a space:

    In the drawing there is a pre-drawn polyline that wraps around the inside of the core area. Zoom to the center of the drawing so that you can see the door located in the waiting area.

    a.  From the Concepts menu, click Spaces ➤ Convert to Spaces.

b. Pick the polyline by picking next to the door located in the waiting area and then pressing ENTER to accept, so that you do not erase the layout geometry.

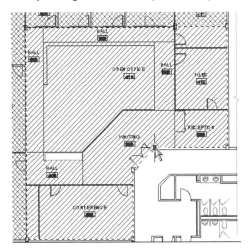

c. On the Style tab, select Zone 3 and then click OK.

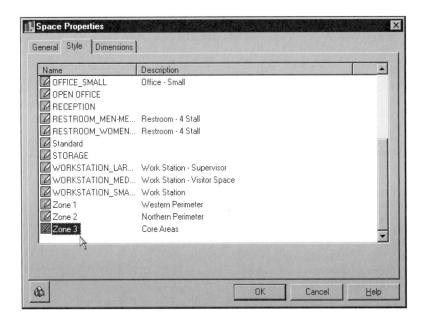

14. Perform a space inquiry to see the total area of Zone 1, Zone 2, and Zone 3:

a. From the Concept Menu, click Spaces ➤ Space Inquiry.
You can also enter **space** at the command line, and enter **Q** for Query.

b. In the Space Information dialog box, click the Space Info Total tab.

You can now view the total square footage for each of the new space styles that you created.

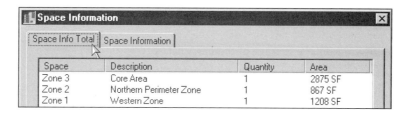

| Space | Description | Quantity | Area |
|-------|-------------|----------|------|
| Zone 3 | Core Area | 1 | 2875 SF |
| Zone 2 | Northern Perimeter Zone | 1 | 867 SF |
| Zone 1 | Western Zone | 1 | 1208 SF |

You now know how to create space styles, add them to an architectural drawing, and perform a space inquiry in order to see the total square footage of an area.

### EXERCISE 2: GENERATING SPACES

In this exercise you add spaces to a drawing where they do not exist using the Generate Spaces command. In the exercise you use the same space styles that you created in exercise 1. The Generate Space function enables you to pick points in the drawing to create spaces similar to creating a hatch.

 **Open:** *Apdx A Creating Spaces.dwg*

1. Select all of the walls in the drawing:

   a. From the Concept Menu, click Spaces ➤ Generate Spaces.
   You can also enter **spaceautogenerate** and the command line.

   b. For Selection Filter select Walls only and then click OK.

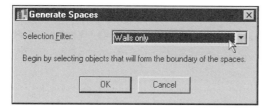

If the drawing that you have been given does not contain walls, then you can select one of the other options.

   c. At the command line, enter **all** and press ENTER twice.

2. Verify that no tags will be added to the spaces:

   a. In the Generate Spaces dialog box, click Tag Settings.

   b. In the Tag Settings dialog box, verify that Add Tag to New Spaces and Add Property Set to New Spaces are deselected.

   c. Click OK to return to the Generate Spaces dialog box.

3. Create spaces:

    a. In the Generate Spaces dialog box, select Zone 1 for style and pick all of the offices along the west wall.

    b. Change the style to Zone 2 and pick all of the offices along the north wall.

    c. Change the style to Zone 3 and pick all of the remaining spaces.

4. View the square footage of the new spaces:

    a. In the Generate Spaces dialog box, click Space Query.

    b. In the Space Information dialog box, click the Space Info Total tab to view the total area of Zones 1, 2, and 3.

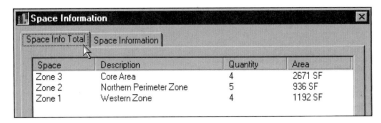

    c. Click OK and then click close.

The difference between the totals shown here and the previous exercise, is that the areas occupied by the walls are not accounted for in the numbers for this lesson.

In Lesson 1, "Calculating Square Footage using Spaces," you completed two exercises: Creating and Querying Spaces and Space Styles, and Generating Spaces. You have now successfully created and added spaces to a drawing. You do this in order to perform a space inquiry so that you can review and share the square footage of the areas contained in your drawing.

## LESSON 2: WORKING WITH AREA AND AREA GROUPS

While Space Objects are relatively simple, the space inquiry may be all that you need. However, Building Systems also provides another set of objects called Areas and Area Groups. While Space object inquiries may be written out into a database, the Area and Area Groups may be used to generate Evaluations, with the square footages written out of the drawing into an Excel spreadsheet.

Area and Area Groups are part of the international extensions. Architectural Desktop 3.0 did not install international extensions by default; however, Building Systems does. You must have international extensions installed to complete the following exercises. You can tell if international extensions are installed by looking at the Documentation pulldown menu. If you see Areas and Area Groups listed on the Documentation menu, you have international extensions installed.

The drawings that you will work with in the following exercises have a display configuration created for the left viewport to show only the area objects, and a display configuration created for the right viewport to show only area groups.

Area and Area Groups have more functionality than you will be learning about in the following exercises. You may also create name definitions to maintain consistent naming of areas across drawings. You can apply an area calculation modifier to areas to have an arithmetic expression applied as the evaluation is written out (such as using areas to create a preliminary cost estimate based on square footage). To learn more about Area and Area Groups, see the Architectural Desktop online Help.

In the following exercises you use pre-created Area and Area Group styles. You are welcome to use Style Manager to copy these styles into your drawings if they are useful to you. You can use DesignCenter to import Area and Area Group layouts by using the customized display configurations created for the following exercises for use in your drawings. If you choose to work with Areas and Area Groups, this is a handy layout to have.

This lesson contains two exercises. Each exercise has a drawing file associated with it. When you complete the exercises in this lesson you will have written out the square footage of an area to a Microsoft® Excel spreadsheet, created areas out of drawing objects, added Area Groups, and attached the Areas to the Area Groups.

To accomplish these tasks you will be using the Area Evaluation dialog box, the Add Areas and Area Groups dialog boxes, and the Attach Areas and Area Groups command.

This lesson contains the following exercises:

- Creating Area Evaluations
- Adding Areas and Area Groups

### EXERCISE 1: CREATING AREA EVALUATIONS

In this exercise you will become familiar with the area evaluation user interface and learn how to write out the square footage of an area to a Microsoft® Excel spreadsheet.

 **Open:** *Apdx A Creating Area Evals.dwg*

1.  View the Area Evaluation dialog box:
    a.  From the Documentation Menu, click Areas ➤ Area Evaluation.
    b.  In the left panel, check the box that is to the left of Zone 2.
    c.  Select Zone 2 and notice in the upper right-hand corner of the dialog box that all of the properties are totaled for all of the areas assigned to the Zone 2 area group.

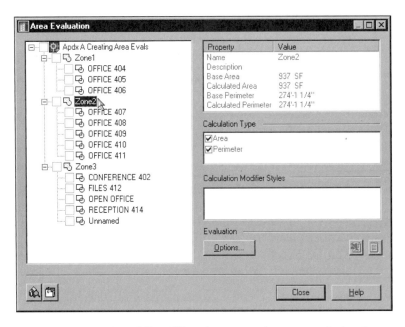

d. In the left panel, select Office 409 and notice in the upper right-hand corner of the dialog box that only the properties of the Office 409 area are provided.

2. Specify options for the area evaluation:

Specifying options for the area evaluations enables you to define what you want to use the square footages for (Areas, Area Groups, or both). Specifying these options also enables you to specify the format that you want to use for your printed area evaluation, which can be either a Microsoft® Excel Spreadsheet, or a Rich Text Format (.RTF) file.

a. Click Options to open the Evaluation Properties dialog box.

b. Click the Evaluation tab and select Area.

c. Verify that Name, Overview Image, Area, Base Area Label, and Base Area Result are all checked. Uncheck all of the other boxes (including Perimeter).

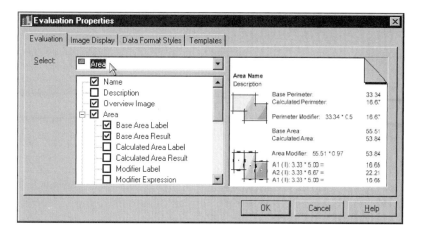

3. Specify the options for the Area Group evaluation:

   a. Under Select, change Area to Area Group.

   b. Verify that Name, Area, Base Area Label, and Base Area Result are all checked. Uncheck all of the other boxes (including Perimeter).

4. Choose a template to use for the printed area evaluation:

   a. Click the Templates tab.

      For the purposes of this exercise, you do not need to use the Image Display tab or the Data Format Styles tab. If you would like to learn more about the options contained on these tabs, click the Help button from each tab.

   b. Under Excel Template File, click the ellipses (...) button.

   c. Browse to *Program Files\Autodesk Architectural Desktop 3\Template\Evaluation Templates*, select the *Area Evaluation Template (1) - Landscape.xlt* file, click on Open, and then click OK.

5. Export the area evaluation to a Microsoft® Excel spreadsheet:

   a. In the Area Evaluations dialog box, click the Excel Spreadsheet icon located in the lower right-hand corner.

   b. Browse to a location where you would like to save your area evaluation spreadsheet.

   c. Enter **Area Evaluation** for File Name and then click Save.

You have completed the task of defining properties for Areas and Area Groups and also exporting an area evaluation to a Microsoft® Excel spreadsheet. In the next exercise you will create and add Areas and Area Groups to the drawing.

### *EXERCISE 2: ADDING AREAS AND AREA GROUPS*

In this exercise you will add area objects into the drawing. Areas look similar to polylines; however, in this exercise you use walls that already exist in the drawing and place areas that resemble a hatch. You could simply add three Areas, one for each zone, and then produce an area evaluation based on those three area zones. However, in this exercise you create the Areas, add Area Groups, and then attach the Areas to the Area Groups. Area Groups enable you to create smaller sets of Areas, and calculate the result. You may also nest Area Groups inside Area Groups. Each level of Area Group that you create gives you the total square footage. This ability is useful when working with a complicated building. This way, you can not only export the square footage of the zone, but also the total square footage of the entire building. If necessary you can also export the square footage of the individual rooms, all in one Excel spreadsheet.

 **Open:** *Apdx A Adding Areas and Groups.dwg*

1.  Select all of the bounding walls in the drawing:

     **Note:** Make sure that the left viewport is active before beginning the following steps.

    a.  From the Documentation Menu, click Areas ➤ Create Area from Object.

    b.  Click Select Bounding Walls.

    c.  At the command line, enter **All** and press ENTER twice.

2.  Specify the office areas:

    a.  In the Create Areas from Objects dialog box, change the Style to Offices.

    b.  Pick once in each of the nine closed offices (three are located on the left, four on the top center, and two on the right).

    A brown diagonal hatch appears over each office space that you pick.

    c.  Zoom into the left viewport to locate the Open Office tag.

3.  Specify the open office area:

    a.  Change the Style to Open Office.

    b.  Pick a point near the Open Office tag.

    A yellow diagonal hatch appears over the open office, the hall, and the waiting area.

    c.  Zoom into the left viewport to locate the conference area and its closet.

4.  Specify the conference area:

    a.  Change the Style to Conference.

    b.  Pick a point in the conference area and pick a point in the closet.

        A black double hatch appears over the conference room and the closet.

    c.  Press ENTER twice to end the command.

5.  Add an area group named Zone 1:

    a.  Click in the right viewport to make it active, and from the Documentation menu, click Area Groups ➤ Add Area Group.

    b.  For Name, enter **Zone 1** and select Zone 1 for Style.

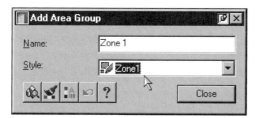

    c.  Pick a point at the lower-left corner of the viewport to place the group.

6.  Add an area group named Zone 2:

    a.  In the Add Area Group dialog box, enter **Zone 2** for Name and select Zone 2 for Style.

    b.  Pick a point at the middle left just above the Zone 1 group to place the Zone 2 group.

7.  Add an area group named Zone 3:

    a.  In the Add Area Group dialog box, enter **Zone 3** for Name and select Zone 3 for Style.

    b.  Pick a point at the top left just above the Zone 2 group to place the Zone 3 group.

    c.  Click Close.

8.  Attach areas to the Zone 1 area group:

    a.  In the right viewport, select the Zone 1 area group, right-click, and click Attach Areas/Area Groups.

    b.  Click in the left viewport, and pick the three office areas along the left wall.

        The offices appear diagonally hatched in red along the left wall of the right viewport.

    c.  Press ENTER.

9. Attach areas to the Zone 2 area group:

   a. In the right viewport, select the Zone 2 area group, right-click, and click Attach Areas/Area Groups.

   b. Click in the left viewport, and pick the five remaining office areas along the top wall.

      The offices appear diagonally hatched in orange along the top wall of the right viewport.

   c. Press ENTER.

10. Attach areas to the Zone 3 area group:

    a. In the right viewport, select the Zone 3 area group, right-click, and click Attach Areas/Area Groups.

    b. Click in the left viewport, and pick the center open office and all of the remaining areas.

       All of the remaining spaces appear diagonally hatched in blue.

    c. Press ENTER.

11. Export the area evaluation to a Microsoft® Excel spreadsheet:

    a. From the Documentation Menu, click Areas ➤ Area Evaluation.

    b. Click the Excel Spreadsheet icon located in the lower right-hand corner.

    c. Browse to a location where you would like to save your area evaluation spreadsheet, enter **Area Evaluation Zones** for Name, click Save, and then click Close.

12. Display the connection from the Zone 1 group to the areas:

    a. In the right viewport, select the Zone 1 area group, right-click and click Edit Area Group Style.

    b. Click the Display Props tab and then click Edit Display Props.

    c. Click the Layer/Color/Linetype Tab and turn on Area Connection Line, and then click OK twice.

In this exercise you added area objects to the drawing that displayed as a hatch. You then added area groups, attached the areas to the area groups, exported the area evaluation to Excel, and displayed connection lines from an area group to its associated areas.

In Lesson 2, "Working with Area and Area Groups," you completed two exercises: Creating Area Evaluations and Adding Areas and Area Groups. You have now successfully created and added areas and area groups to a drawing. You do this in order to create an area evaluation Excel spreadsheet so that you can start the design and sizing of your mechanical system, and also review and share the square footage of the areas contained in your drawing.

# KEY CONCEPTS: CALCULATING FLOOR AREAS

You can calculate the floor area of a building using spaces, or Areas and Area Groups.

A space object is comprised of a floor and a ceiling.

Architects use spaces to represent rooms.

Spaces have a built-in square footage calculator.

Spaces are style-based objects.

Area and Area Groups are part of Architectural Desktop's international extensions.

You can create areas out of drawing objects, such as bounding walls.

Like spaces, Areas and Area Groups are also style-based objects.

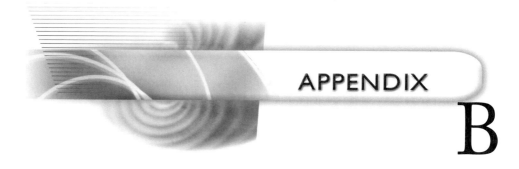

# Exploding Your Drawings

After you complete the exercise in this chapter, you should be able to:

- understand proxy graphics.
- understand object enablers.
- know how and why you should explode your drawings.

## USING OBJECT EXPLODE

At some point during your work flow process you will need to give your drawings in .DWG format to somebody else outside of your office, or perhaps even a department within your office that does not work with Autodesk Building Systems. However, drawings created using Building Systems can only be manipulated using Building Systems. This appendix addresses the need to change drawings using AutoCAD or Architectural Desktop without having Building Systems installed. The reason that AutoCAD or Architectural Desktop cannot directly change drawing objects that were created in Building Systems are because objects contain a set of instructions that tell your graphics card how to display themselves in your drawing. That set of instructions is termed proxy graphics. How the other members of your team are able to see and control proxy graphics can be determined by answering the following questions:

From your end: What is the proxy graphics setting specified as in your drawing? Are the graphics that make up the objects stored in the drawing or not?

 **Note:** The default the proxy graphic setting is zero.

From the receiving end: What version of AutoCAD or Architectural Desktop are they using? What object enablers do they have installed?

## UNDERSTANDING PROXY GRAPHICS

Proxy graphics are the geometry that you see when you look at a 1-line or 2-line representation of a duct segment. A duct segment is comprised of just two pick points, just like an AutoCAD line, a duct segment has a start and an end point. All AutoCAD knows when displaying a duct segment is that there is a vector with a length. It is Building Systems set of instructions to the graphic card that displays the duct segment as 1-line, 2-line or three-dimensional (3D). If proxy graphics are set to zero (the default) then the geometry that you see is not saved in the drawing, just the information to display it the next time that it is opened in Building Systems. So when the drawing is opened in AutoCAD, you are only given a bunch of bounding boxes with the name of the object set down. The instructions for displaying the object are there, but there is no mechanism for interpreting those instructions. If proxy graphics are set to one, then the last graphic that you see when you saved and exited the drawing is saved as geometry that an AutoCAD user can see also.

Remember, proxy graphics only respect the last graphics that were saved. For example, if you are looking at a 1-line view of the system, then the AutoCAD user will see the 1-line view of the system when they open the drawing, similarly if the last representation is 3D, then no matter how the AutoCAD user looks at the system, it will be in 3D. In either of these situations, the AutoCAD user will not be able to change anything about the objects. The AutoCAD user will not be able to move, stretch, change the layer of the object or modify the object in any way because they are only proxy graphics, and AutoCAD cannot read any of the 'built in' values of the object.

## USING OBJECT ENABLERS

Object enablers are free downloads supplied by Autodesk to work around the inability to work with objects that were created in different AutoCAD products or releases. If you give your drawing to another team member using ACAD 2002, they can download the proper object enabler and the object enabler does the trick of interpreting the additional information held by the objects. Objects enablers allow the AutoCAD user to change properties such as the layer and color, and are also supplied with the ability to stretch and move Building Systems objects.

Object enablers are not backwards compatible, in other words there is no object enabler for somebody using AutoCAD R14 or 2000 available that can read a Building Systems object, because Building Systems was released after AutoCAD R14 and 2000.

Using object enabler will not always be the answer to the question, "How do I give this file to other people who must work with my drawings?" Sometimes the only answer is to explode the objects. This appendix is included to give you some of the finer points of exploding your drawings. AecObjExplode is a command that does just that. Taking any object down to base AutoCAD entities such as lines, arcs, and faces that anyone with AutoCAD can use. Unfortunately, MvParts do not always work the same as other objects, therefore, this appendix presents a work around that gives you a two-dimensional (2D) plan of linework which is standard AutoCAD lines at elevation Z=0, where all of the linework is on the proper layer, and the properties are ByLayer.

## LESSON OBJECTIVES

This appendix contains one lesson. This lesson contains only one exercise. This exercise has a corresponding drawing file that is located on the CD included with this book.

Lessons in this appendix include:

- **Lesson 1: Using Object Explode**

    You open a drawing that contains AEC objects, and then reduce the drawing to standard AutoCAD entities by using object explode.

## LESSON 1: USING OBJECT EXPLODE

The associated drawing file for the following exercise contains AEC objects that cannot be read or manipulated by others who may not have Building Systems or Architectural Desktop. This lesson uses the object explode command to make the drawings readable for others who do not have the same programs or program release numbers installed.

This lesson contains the following exercise:

- Exploding your Drawing

### EXERCISE 1: EXPLODING YOUR DRAWING

Object explode takes whatever view is current and explodes the objects down to standard AutoCAD entities. To do this it will look at the display representations settings, and use those settings to find what layers and colors to assign to the subcomponents. In this exercise you will see how the program does this.

**Open:** *Appendix_Explode.dwg*

1. Review the options on the Explode AEC Objects dialog box:
   a. Click in the lower left viewport to make it active.
   b. From the Desktop menu, click Utilities ➤ Explode AEC Objects.
   c. Select the following options and then click OK.

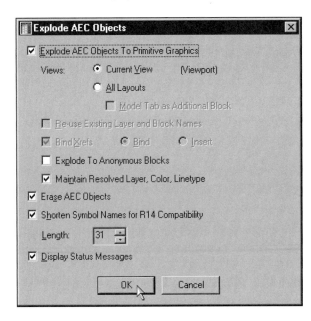

| You Should Select... | Because... |
|---|---|
| Explode AEC Objects to Primitive Graphics | the objects are reduced to basic AutoCAD lines, arcs, and circles that can be read by others who do not have Building Systems. |
| Current View | otherwise the program creates a block for each viewport in the drawing all stacked on top on one another. |
| Maintain Resolved Layer, Color, Linetype | this prompts the program to look at the layer, color and linetype of the object and respect those properties when the object is exploded. Otherwise, all of the sub-objects are reduced to layer zero and ByBlock. |
| Erase AEC objects | You want to 'get rid' of AEC objects so that AutoCAD users can view and work with the objects. |
| Shorten Symbol Names for R14 Compatibility | This option enables anyone with AutoCAD R14 to work with the drawing, otherwise the symbol names are too long. Do not shorten names below 22 characters. |

All of the viewports now show the same thing because what you are seeing are standard AutoCAD blocks that have no 'instructions' telling the display to show themselves any other way. Notice that all of the MvParts are exploded down to layer zero. This behavior happens only in the 1-line view.

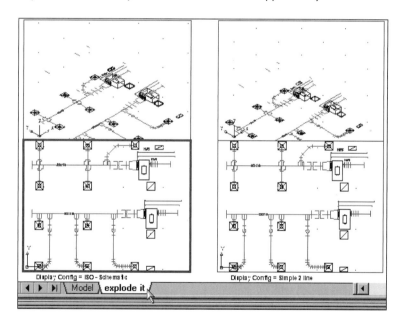

2. Explode your drawing so that the layer of MvParts remain unchanged:

   a. Undo the last command by entering **undo**, press ENTER twice and then click in the lower right viewport to make it active.

b.  From the Desktop menu, click Utilities ➤ Explode AEC Objects and select the same options that you did in step 1, except this time also select Explode To Anonymous Blocks.

If Explode To Anonymous Blocks is not checked then all MvParts are exploded down to layer zero.

c.  Click OK and view blocks in your drawing.

Now because you selected the Explode to Anonymous Blocks option, each object becomes an individual block. If you explode a duct and an MvPart, for example one of the return diffusers, you see why you are not quite done. The exploded duct explodes to linework on layer M-Duct-Supply, which is good, however it also assigned the color 32, which is taken from the layer, which may not be the result you are looking for.

If you are giving this drawing to someone who just needs to turn layers on and off, and you have control over all of the plots of this drawing your job may be done.

If however you are giving this drawing to anyone who needs control over the color and the linework, then you should proceed with the next set of steps. You may need to undo the last command by entering **undo** and then pressing ENTER twice.

## Create a 2D ByLayer Drawing

3.  Create a 2D block:

a.  Click the lower right viewport to make it active and from the Desktop menu, click Utilities ➤ Hidden Line Projection.

b.  Enter **all** at the command line and press ENTER twice to select all of the objects.

c.  Pick a point in the lower right hand corner of the viewport as the insertion point and press ENTER to accept 'insert in Plan view Y' for the default.

4.  Erase the extra linework in the drawing:

a.  Erase all of the original linework except for the reference circle at the upper left of the viewport.

b.  Zoom out until you can see both of the circles with the cross in them (these are just reference points).

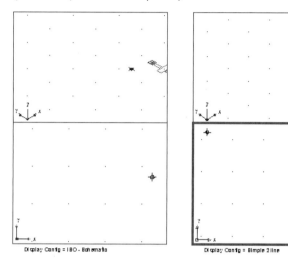

c.  Using these reference circles, move the block created by the hidden line projection onto the original location of the ducts.

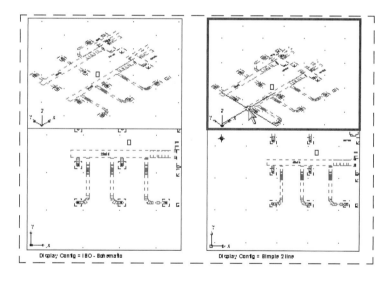

5.  Change the Linework to ByLayer:

    a.  Explode the block in the drawing by selecting the block and then from the Modify menu clicking Explode.

    b.  Select all of the linework in the drawing with a crossing or window selection and on the Object Properties toolbar select ByLayer.

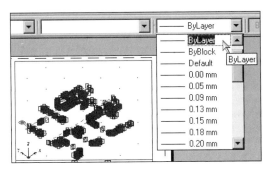

6.  Purge the drawing:

    a.  From the File menu, click Drawing Utilities ➤ Purge.

    b.  Select all of the items in the drawing and click the purge button.

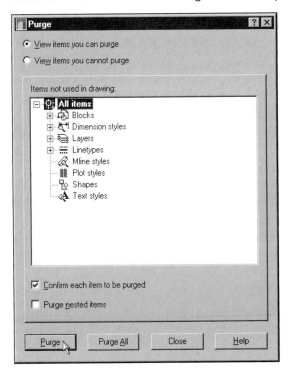

c. Click Yes to All in the Confirm Purge dialog box and then click Purge All in the Purge dialog box.

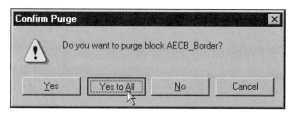

d. From the File menu click Save As, and then select AutoCAD R14/LT 98/LT 97 Drawing and then close the drawing.

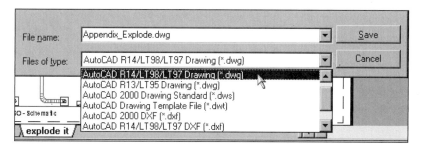

You perform a Save As so that you can give this drawing to an AutoCAD R14 or AutoCAD LT user.

 **Caution:** Avoid launching any Building Systems commands, as this will cause more proxygraphics to be created in the drawing, and whoever receives this drawing will get the proxygraphics error message when it is opened.

In this exercise you exploded a drawing so that AutoCAD users could view and change the drawing properties.

## KEY CONCEPTS: EXPLODING YOUR DRAWINGS

AEC objects contain a set of instructions that tell your graphics card how to display themselves in your drawing. That set of instructions is termed proxy graphics.

Proxy graphics are the geometry that you see when you look at a 1-line or 2-line representation of a duct segment.

If proxy graphics are set to zero (the default) then the geometry that you see is not saved in the drawing, just the information to display it the next time that it is opened in Building Systems.

If proxy graphics are set to one, then the last graphic that you see when you saved and exited the drawing is saved as geometry that an AutoCAD user can see also.

Object enablers are free downloads supplied by Autodesk to work around the inability to work with objects that were created in different AutoCAD products or releases.

Object enablers are not backwards compatible.

# INDEX

# LICENSE AGREEMENT FOR AUTODESK PRESS
## A Thomson Learning Company

### Educational Software/Data

You the customer, and Autodesk Press incur certain benefits, rights, and obligations to each other when you open this package and use the software/data it contains. BE SURE YOU READ THE LICENSE AGREE-MENT CAREFULLY, SINCE BY USING THE SOFTWARE/DATA YOU INDICATE YOU HAVE READ, UNDERSTOOD, AND ACCEPTED THE TERMS OF THIS AGREEMENT.

### Your rights:

1. You enjoy a non-exclusive license to use the enclosed software/data on a single microcomputer that is not part of a network or multi-machine system in consideration for payment of the required license fee, (which may be included in the purchase price of an accompanying print component), or receipt of this software/data, and your acceptance of the terms and conditions of this agreement.

2. You own the media on which the software/data is recorded, but you acknowledge that you do not own the software/data recorded on them. You also acknowledge that the software/data is furnished "as is," and contains copyrighted and/or proprietary and confidential information of Autodesk Press or its licensors.

3. If you do not accept the terms of this license agreement you may return the media within 30 days. However, you may not use the software during this period.

### There are limitations on your rights:

1. You may not copy or print the software/data for any reason whatsoever, except to install it on a hard drive on a single microcomputer and to make one archival copy, unless copying or printing is expressly permitted in writing or statements recorded on the diskette(s).

2. You may not revise, translate, convert, disassemble or otherwise reverse engineer the software/data except that you may add to or rearrange any data recorded on the media as part of the normal use of the software/data.

3. You may not sell, license, lease, rent, loan, or otherwise distribute or network the software/data except that you may give the software/data to a student or and instructor for use at school or, temporarily at home.

Should you fail to abide by the Copyright Law of the United States as it applies to this software/data your license to use it will become invalid. You agree to erase or otherwise destroy the software/data immediately after receiving note of Autodesk Press' termination of this agreement for violation of its provisions.

Autodesk Press gives you a LIMITED WARRANTY covering the enclosed software/data. The LIMITED WARRANTY can be found in this product and/or the instructor's manual that accompanies it.

This license is the entire agreement between you and Autodesk Press interpreted and enforced under New York law.

### Limited Warranty

Autodesk Press warrants to the original licensee/ purchaser of this copy of microcomputer software/ data and the media on which it is recorded that the media will be free from defects in material and workmanship for ninety (90) days from the date of original purchase. All implied warranties are limited in duration to this ninety (90) day period. THEREAFTER, ANY IMPLIED WARRANTIES, INCLUDING IMPLIED WARRANTIES OF MERCHANTABILITY AND FITNESS FOR A PARTICULAR PURPOSE ARE EXCLUDED. THIS WARRANTY IS IN LIEU OF ALL OTHER WARRANTIES, WHETHER ORAL OR WRITTEN, EXPRESSED OR IMPLIED.

If you believe the media is defective, please return it during the ninety day period to the address shown below. A defective diskette will be replaced without charge provided that it has not been subjected to misuse or damage.

This warranty does not extend to the software or information recorded on the media. The software and information are provided "AS IS." Any statements made about the utility of the software or information are not to be considered as express or implied warranties. Delmar will not be liable for incidental or consequential damages of any kind incurred by you, the consumer, or any other user.

Some states do not allow the exclusion or limitation of incidental or consequential damages, or limitations on the duration of implied warranties, so the above limitation or exclusion may not apply to you. This warranty gives you specific legal rights, and you may also have other rights which vary from state to state. Address all correspondence to:

AutodeskPress
Executive Woods
5 Maxwell Drive
Clifton Park, New York 12065-2919